LONESOME GEORGE

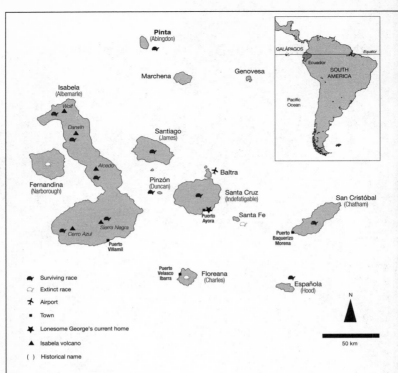

The Galápagos Islands

The Galápagos archipelago is made up of around a dozen sizeable islands and more than 100 rocky outcrops. This volcanic chain sits in the Pacific Ocean about 1000 km off the west coast of South America and has emerged from the sea floor within the last few million years.

Throughout their history, these islands have been named and renamed, some being christened more than 10 times. As a general rule, a suite of (quint)essentially British names survived into the 20th century, when the current Ecuadorian nomenclature achieved the ascendancy. Lonesome George uses these modern names wherever possible to avoid confusion. In chapters citing historical texts, some of the old names appear, with the contemporary equivalent given in brackets. For example Lonesome George's home, today called Pinta, was once known as Abingdon Island.

LONESOME GEORGE

The Life and Loves of a Conservation Icon

Henry Nicholls

Macmillan
London   New York   Melbourne   Hong Kong

First published 2006 by
Macmillan
Houndmills, Basingstoke, Hampshire RG21 6XS and
175 Fifth Avenue, New York, N.Y. 10010
Companies and representatives throughout the world

ISBN-13: 978–1–4039–4576–1
ISBN-10: 1–4039–4576–4

This book is printed on paper suitable for recycling and made from fully
managed and sustained forest sources.

A catalogue record for this book is available from the British Library.

A catalog record for this book is available from the Library of Congress.

10 9 8 7 6 5 4 3 2 1
15 14 13 12 11 10 09 08 07 06

Printed and bound in China

To Harry

CONTENTS

LIST OF FIGURES

ACKNOWLEDGEMENTS

This project came about because of Fausto Arellano, my naturalist guide in the Galápagos; his wit and flair inspired me to explore Lonesome George's wonderful story further. Peter Aldhous and Nicola Jones pulled together my early foray into George's world for a feature published in *Nature* in June 2003. The book was an obvious next step. Thanks to my editor at Macmillan Science Sara Abdulla for listening to my verbal pitch, encouraging me to put together a proposal, taking it on and guiding me from start to finish. It's been super fun.

Everyone I've interviewed for the book, without exception, has been immensely cooperative. I appreciate the time and thought they gave in answering my questions.

Lengthy discussions and correspondence with Linda Cayot and Peter Pritchard enriched almost every angle of the book. In addition to their opinions and recollections, several others helped me piece together George's narrative: Thomas Fritts, Manuel Cruz, Ole Hamann, Ole Seberg, Peter Kramer, Sveva Grigioni, Gisela von Hegel, Derek Green, Rob Gradstein, Joe Flanagan, Olav Oftedal, Eliott Jacobson, Howard Snell, Roslyn Cameron and Víctor Carrión. I benefitted hugely from conversations with historians of science Edward J. Larson, Frank Sulloway, Janet Browne, John Woram, Jordan Goodman, Paul White, Joy Harvey, John van Wyhe, Kristin Johnson and John Wills. Tracking details of zoo and museum specimens was down to help from Colin McCarthy, Douglas Russell, George Zug, Steve Johnson, Geoffrey Swinney and Petr Velenský. The tales of rediscovery of extinct species are courtesy of interviews with Peter Zahler, Robert Dowler,

Justin Gerlach and Pamela Rasmussen. Thanks to the Yale geneticists Adalgisa Caccone, Jeffrey Powell and Michael Russello, to ancient DNA experts Alan Cooper, Svante Pääbo and Tom Gilbert and to Michel Milinkovitch, David Kizirian and Edward Louis, all of whom gave life to the stories hidden in DNA. Nigel Leader-Williams and Matt Walpole talked to me about flagship species. Andrew Balmford, Claudio Sillero-Zubiri, Paul Ferraro, Michael LeMaster, Scott Keogh, Kim Parsons, Margarida Fernandes, Samuel Wasser, Tony Juniper, Richard Lewis and Farah Ishtiaq let me pick their brains over conservation beyond the Galápagos. I learned about the ins and outs of the sea cucumber crisis from Chantal Blanton, Jim Pinson, Verónica Toral-Granda, Chantal Conand and Graham Edgar. I got up to date about current conservation initiatives in the islands from Karl Campbell, Josh Donlan, Donna Harris, Johannah Barry and Graham Watkins. My appreciation of the difficulties in collecting semen from elephants comes from a conversation with Thomas Hilde-brant; an understanding of electroejaculating reptiles from a chat with Carrol Platz Jr; an introduction to reptilian sperm storage from Daniel Gist; and comprehension of the complex-ities of assisted reproductive technologies from Bill Holt, Valentine Lance, Tim Birkhead and Beatrix Schramm. The following helped me explore the futuristic worlds of chimeras and cloning: Ian Wilmut, Pasqualino Loi, Grazyna Ptak, Salvatore Naitana, James Petitte, Oliver Ryder, Gordon Woods and Don Jacklin. The thoughts of William White, Kirsten Berry, Larry Agenbroad, Peter Tallack, Tom Tyler, Matthew James, Greg Moss, Anthony and Setitia Simmonds, Roy Easson, Mike Spurgin, Bob Langton and Don Freeman also fed into the book.

Polly Tucker and the staff of the library at the Natural History Museum were tremendously helpful in allowing access to the Rothschild Collection and many other fascinating sources; Kate Jarvis dredged up logbooks from the National

Maritime Museum; and it's been a pleasure working with all those at the Galápagos Conservation Trust, especially Leonor Stjepic, Abigail Rowley and Catherine Armstrong. I've also valued discussions with trustee Nigel Sitwell and vice-presidents Julian Fitter, Sarah Darwin and Godfrey Merlen.

I'd like to thank everyone who has provided illustrations. Detailed credits are listed at the back. In particular, I'm indebted to John Woram for letting me reproduce maps from his enchanting website (www.galapagos.to) from Cowley's visit, the *Beagle* voyage and the sketch of the Abingdon tortoise that appears in the 11th edition of Darwin's *Journal of Researches*.

Many others have helped me through this adventure, especially my family John, Stella, Mary, Tom, Ana, Pablo and Alvaro Nicholls, Mark Ruddy, Hugh and Sheila Stirling, James and Hazel Mason and friends Zaid Al-Zaidy, Kate Moorcroft, Matthew and Marisa Lea, Pia Sarma, Rufus Grantham, Jimmy Carr, Melvin Carvalho, Laura Cook, Jonathan Duffy, Arthur Wadsworth, Julian Ogilvie, Matthew Thorne, Daniel Price, Gina Fullerlove, Caroline Tullis, Mark Wilson, Tommaso Pizzari, Martin Fowlie, Phil Mitchell, Ben Keatinge, James Samson, Bea Perks, Helen Dell, Ruth Jordan, Laura Spinney, Pete Moore, Catriona MacCallum, Colin Tudge, Darren Sharpe, Max Benitz and the Celeriac XI.

Linda Cayot, Graham Watkins, John Van Wyhe, Peter Pritchard, Pia Sarma, Tim Birkhead, Mark Wilson and Tom Nicholls deserve an additional mention for their valuable feedback on earlier versions of the book.

My son Harry arrived in the midst of this project and has been a limitless source of delightful distraction throughout. I wouldn't have done this without him or my wife Charlotte, who has supported me intellectually and emotionally every slow step of the way. Thank you.

HENRY NICHOLLS

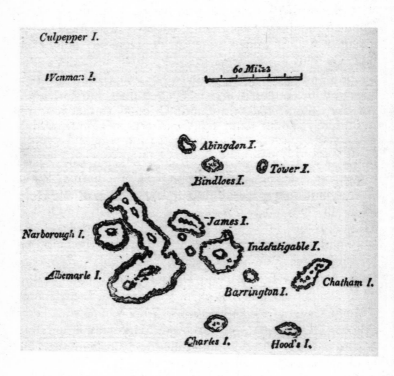

Figure P.1 The Galápagos archipelago

A CONSERVATION ICON

The aeroplane pushes out over the Pacific Ocean towards the Galápagos. Sitting near the back of the plane, I squint through parting clouds into the shimmering sea for a first glimpse of the islands – the naturalists' Mecca.

My well-thumbed second edition of Charles Darwin's *Journal of Researches* lies open in my lap at Chapter 17. I've spent the flight out rereading Darwin's account of his five-week journey through the Galápagos in September and October 1835 that sparked his thoughts about evolution. I compare the sketch of the main islands with the landscape now twinkling below, trying to work out which one we're passing over. I'm struck by the remarkable continuum of colour from the lowest to the highest point of each island: barren brown on the rocky shores bleeding to tropical green near the top of each volcano. The plane goes into descent. '*Diez minutos para el aterrizaje*' – 'Ten minutes to landing.'

Ten years before me, a 26-year-old Swiss zoology graduate made the same two-and-a-half-hour journey from the mountainous backdrop of Quito to the stunning Galápagos Islands, some 1000 km off Ecuador's coast. It was April 1993 and Sveva Grigioni was on her way to work at the Charles Darwin Research Station (CDRS), the international body responsible for science in the archipelago, based in its main town Puerto Ayora on the central island of Santa Cruz.

The CDRS gave Grigioni two research options: she could help with some gecko work or single-handedly attempt to persuade Lonesome George, the research station's resident celebrity, to take an interest in females.

Lonesome George is the world's most famous reptile. A legend in his own long lifetime, George is thought to be the only giant tortoise from the isolated island of Pinta to have survived centuries of buccaneers and whalers in search of a square meal. Before he was discovered, the Pinta tortoise was assumed extinct. George brought hope – and thousands of tourists. But as each year comes and goes, it looks more and more like George is the only one of his kind left on earth – a symbol of the devastation man has wrought to the natural world in the Galápagos and beyond.

For Grigioni, there was no question: she chose the taciturn tortoise. 'I had the feeling that I was about to do something very important', she says.

We drop gently out of the sky. After a week in the mists and rain of Quito, it's exciting to be surrounded by sea. The wheels unfold with a clunk as the plane nears Baltra, the small central island kitted out with a runway by the US military in the Second World War. Since the 1970s, Baltra has served as the entry point for most visitors to this, the best-preserved tropical archipelago in the world.

I flip to the frontispiece of the book; from behind a crisp film of transparent paper stare the penetrating eyes of an ageing Darwin, his long, white, wise man's beard framing the lapels of a dark cloak that flows off the bottom of the page.

This is a journey I've been planning since my days as a zoology undergraduate at Cambridge. I want to see for myself the place that drew Darwin towards one of the most influential ideas in the history of science – his theory of evolution by natural selection. I want to sail into the same coves and set foot on the same beaches. I also want, I confess, to find a quiet spot beneath a cactus and read aloud from his writings, one sentence in particular: 'The natural history of these islands is eminently curious.' I love that. More than anything,

Figure P.2 Charles Darwin, from the frontispiece to his *Journal of Researches*

this is a journey to see these same eminently curious creatures: the archipelago's extraordinary array of finches, its amphibious, algae-eating marine iguanas and of course its arcane giant tortoises.

A few days later, as we sail around the islands, a chirpy guide is regaling me and my fellow tourists about one tortoise in particular – Lonesome George. When he begins telling us about the love affair between this celebrated beast and a beautiful Swiss girl, it's clear this is one of his set pieces. He keeps a straight face, but his eyes sparkle as he describes her work to his startled audience.

If you tell an anecdote often enough, you quickly hone your routine. You embellish what works and drop what doesn't. I can't fault his delivery, but reflect that this yarn and others he tells us about George are but playful shadows of something much richer. It strikes me that this tortoise has another tale that needs to be told.

So here it is. The story of a creature that touches all who see and hear about him, an animal whose plight embodies the practical, philosophical and ethical challenges of preserving our fragile planet. The story of a conservation icon.

Chapter 1
DISCOVERY

On 1 December 1971, American snail biologist Joseph Vagvolgyi and his wife Maria were on Pinta when they came face to face with a giant tortoise. 'The tortoise was walking slowly when we first encountered him, but withdrew into his shell with a loud hiss as we moved closer to take his picture', Vagvolgyi recalled. 'He soon relaxed, and resumed his walk.' Vagvolgyi took a photograph and returned to the undergrowth and his search for snails. Neither he nor his wife realized the immense significance of their encounter.

Figure 1.1 The first photograph of Lonesome George, taken by Joseph Vagvolgyi on 1 December 1971

To most other visitors, the sight of a tortoise on Pinta would have been incredible. As far as everyone knew (except, it seems, the Vagvolgyis), there were no tortoises left on the island. Two centuries of exploitation at the hands of buccaneers and whalers had taken their toll; the last tortoise seen on Pinta was collected by scientists in 1906.

In March 1972, Joseph and Maria hosted dinner in the small free-standing cottage they rented in Puerto Ayora on Santa Cruz. Maria was in the tiny dining room, barely big enough for a table and four chairs, when the doorbell chimed. Joseph ushered in Peter Pritchard, his wife Sibille and their baby son.

Pritchard is a world authority on sea turtles. He grew up in Belfast but has lived in Florida since 1965 where he founded the Chelonian Research Institute. Shortly after the birth of his first son in February 1972, Pritchard's young family flew to the Galápagos from Florida to catch the end of the sea turtle nesting season. Soon after their arrival they received the invitation to dine with the Vagvolgyis.

As often happens when scientists get together in the Galápagos, the conversation turned to giant tortoises. Pritchard found himself telling a captive audience about the two main types in the archipelago: the domed variety and those with shells shaped like saddles. Sibille had heard it all before. Still, she enjoyed seeing her husband in his element. With her son asleep in the bedroom next door, she was relaxed and happy.

When Pritchard described the Pinta tortoise as a saddle-back, Vagvolgyi piped up: 'The tortoise we saw on Pinta did not have a particularly saddlebacked shell.' Pritchard was stunned. 'I practically lost my teeth', is how he remembers it.

Peter Pritchard has been passionate about turtles and tortoises for as long as he can remember: 'I was a pretty weird kid. I tended to be attracted to things that no one else was interested in.' His parents fuelled his obsession. They gave him the *Handbook of Turtles* by Archie Carr, who later guided Pritchard through a doctorate at the University of Florida. For his 16th birthday, he got a copy of *The Gigantic Land-Tortoises (Living and Extinct) in the Collection of the British Museum*, written in 1877 by the museum's curator of reptiles Albert Günther. This 98-page monograph is full of wonderful engravings. One of these – 'a peculiar beast whose shell looked more like a wrinkled, leathery mantle than a true carapace, and whose long, stout neck, raised straight up, carried a tiny head with expressionless eyes' – made a particularly strong impression on the youthful reptile geek. 'I resolved', he wrote in *Natural History Magazine* in 1977, 'that someday I would lead an expedition to try and find survivors of this prehistoric-looking creature.'

Figure 1.2 The 'antediluvian' Pinta tortoise engravings that inspired Peter Pritchard

So here was Pritchard, sitting at a cramped dinner table on an unseasonably cool Galápagos night with a snail fancier who claimed to have seen his reptilian Moby Dick. Unable to contain his excitement, Pritchard bombarded his poor host with questions. When had he seen this tortoise? 1st December. Where was it? On a southerly slope of the volcano. At what kind of altitude? Around 200 m. What size? Not so big. What was it doing? Walking. Had he taken any photos? Yes. How many? One. From what angle? Side on. Could he mail a copy of the photo once it was developed? Yes.

It was a boyhood dream come true. That night, Pritchard didn't sleep. How fantastic it would be, he thought, to track down Vagvolgyi's tortoise.

There are some cracking tales of the rediscovery of supposedly extinct creatures. The woolly flying squirrel – the world's largest – was known only from a few skins collected in late 19th-century Pakistan and from a photograph snapped in 1924 of a British colonel leading one by a rope. For most of the 20th century, naturalists assumed that this bizarre mammal was no more. Then in the summer of 1994, the squirrel was rediscovered in the Northern Areas region in Pakistan by Peter Zahler, an American zoologist working for the Wildlife Conservation Society.

All the valleys that Zahler combed and all the traps he set had drawn a blank, when two men dropped into his camp and offered to supply a living squirrel in exchange for around $40. Holding out little hope, Zahler agreed and handed over a big cloth bag. Within a couple of hours, the men were back with something of about the right size wriggling in the bag. 'It's a woolly flying squirrel', announced Zahler with surprising sangfroid as the animal dropped into a waiting cage. The creature's droppings are reputed to have aphrodisiac properties and the hired hunters were in the business of collecting them. Zahler now has circumstantial evidence of probably a few thousand woolly flying squirrels living near Pakistan's borders with Afghanistan and China.

Another wonderful rediscovery is that of the rice rat of the Galápagos island of Santiago. Until 1997, the rodent was known only from specimens collected in 1906 by the scientists of the California Academy of Sciences in San Francisco. When they eventually got round to studying these in 1932, it looked like they'd found a new species, *Nesoryzomys swarthi*. Unfortunately, by then the presence of non-native black rats on Santiago and the complete absence of any further sightings of the rice rat led most to assume it had gone extinct. A skull turned up in 1965 but extensive field surveys from the 1970s onwards drew a complete blank.

Then in 1997, Robert Dowler, a mammal biologist at Angelo State University in Texas, landed on the north shore of Santiago with his graduate students Darin Carroll and Cody Edwards. They planned to set out early next morning for the highest point on the island, where they thought they stood the best chance of finding rice rats. Setting up camp near the shore, Dowler suggested they put out a few traps just to see what was there.

The next morning the trio was astonished to find 25 Santiago rice rats in the traps. Without a permit to take a sample specimen, they had to let them go. Excited, they hiked up into the highlands, where they managed to make radio contact with officials of the Galápagos National Park Service (GNPS) back in Puerto Ayora who agreed they could collect up to five specimens if they could find any more. Trapping in the highlands yielded nothing. On their return to the shore, they set the traps again. The next morning there were 49 rice rats. Dowler set loose 44 and carried off five for further study.

Back in late 1971, when Vagvolgyi clocked the tortoise on Pinta, German-born conservationist Peter Kramer had just begun his stint as the director of the Charles Darwin Research Station (CDRS). News of the tortoise sighting came as a shock. Kramer had spent three weeks on Pinta studying finches nearly nine years earlier, in March 1963. 'I remember seeing bleached bones of tortoises but nothing alive', he says. 'I was absolutely convinced that it had gone.'

In 1968, the Ecuadorian government created the nucleus of a park service under the control of the Ministry of Agriculture's Forestry Service. At first, the fledgling GNPS shared offices with the research station in Puerto Ayora. This established an alliance that continues today, with scientists and park wardens working closely to protect the unique archipelago. So in 1971 Kramer and his CDRS scientists got together with their GNPS colleagues to discuss what, if anything, should be done about Vagvolgyi's tortoise. They decided that an impending hunting trip to Pinta to cull the thousands of introduced goats destroying the island's plants offered the perfect opportunity to locate the mysterious reptile.

Around the time the Pritchards dined with the Vagvolgyis – early March 1972 – a boat bound for Pinta left the dock of the CDRS and motored out into the bay. Aboard were eleven park wardens and a young zoologist called Manuel Cruz. Cruz was in his last year of studies at the Faculty of Natural Sciences at the University of Guayaquil on mainland Ecuador. He was tasked with finding out what plants the goats were eating by opening up their stomachs to extract half-digested vegetation.

The goat hunters broke their 200-km, two-day voyage to Pinta on Santiago and reached their final destination the following evening. Their boat motored off to return in due course, leaving them on the island. They set up a camp at its southernmost tip. Teenage warden Francisco Castañeda was assigned to help Cruz. Each day, they set off together, armed with a knife, plant press, scales, altimeter, a few plant books and rifles to pick off goats and work out their feeding preferences.

On 20 March, Cruz and Castañeda were skirting the western slopes of the volcano, when they caught sight of something about 60 m ahead. 'We both thought it was a goat and taking aim with our rifles we walked closer until we saw that it was actually a tortoise!' Cruz later wrote in the islands' scientific journal *Noticias de Galápagos*.

The tortoise was surrounded by rocks and feeding from a tree. Cruz handed a camera to Castañeda and clambered over a few boulders to be photographed beside the tortoise. Castañeda then stayed with the animal, while Cruz ran back to the camp with news of their discovery. He took off his shirt and left it hanging like a flag from the tree to help him retrace his steps.

Back at camp, nobody believed the youngster. The absence of Castañeda, however, suggested that something was up. So one warden, Camilo Calapucha, agreed to follow Cruz up the volcano to check out his story. When Cruz caught sight of his shirt in the tree, he raced on ahead. Castañeda, sitting on a large chunk of lava, looked up. To Calapucha's surprise, there was the tortoise, exactly as Cruz had described it.

Calapucha raced back to camp to get help. He returned with all the wardens he could find, some machetes, rope and another camera. One warden, Oswaldo Chapi, took some photos while the others cut down branches from which to sling the tortoise. 'The swinging of the tortoise made it very difficult for us to walk over the lava', Cruz recalled. On two occasions, a branch snapped under its weight. 'It was a horrible trip!'

By the time they made it back to camp, it was the afternoon. The tortoise caused a real stir. It was big and seemed to be in fine condition. Once righted, unleashed from its shackles and set down on the ground, it began to march away from its captors. To prevent its escape, the wardens looped a rope around one of its rear feet and tethered it to a massive cactus.

●

Meanwhile, out at sea, another boat was on its way to Pinta, chartered by Cruz's supervisor Danish botanist Ole Hamann. He had several people in tow, including his wife Michelle, a German iguana specialist Dagmar Werner, a graduate student and a couple of assistants. Also on board were Peter and Sibille Pritchard and their baby son.

Pritchard had got wind of Hamann's expedition and eagerly arranged to share the charter. Nobody had ever studied sea turtles on Pinta and this was a perfect opportunity to survey its beaches for signs of nesting. At least that was his official line. Secretly, Pritchard wanted to be a part of the tortoise adventure and even hoped to be the one to find Vagvolgyi's beast. 'I was much more excited about the possibility of seeing a tortoise on Pinta than a sea turtle', he admits.

The ship's radio started up. A warden on Pinta was contacting the approaching vessel. Pritchard looked on as Hamann took the receiver and began to speak in fluent Spanish. Something was up. At the end of a lengthy and animated conversation, Hamann turned to Pritchard. 'They have found a tortoise, a large male', he beamed.

For an instant, Pritchard was overcome by disappointment. Someone had beaten him to the discovery. This feeling soon faded. Vagvolgyi said he'd seen a smallish tortoise on Pinta. So the large male that the wardens had just found was clearly a different tortoise, he reasoned. Vagvolgyi's beast was still out there to be discovered. What's more, it's smaller size meant that it might be a female. 'There was still hope that a potential breeding pair existed', he later wrote.

The last few hours of the voyage seemed to take forever. By late afternoon, Pinta came into view. As they drew closer, they could make out the camp on the beach. Pritchard scoured the scene through binoculars for signs of the tortoise.

He jumped out onto the lone sandy beach of Pinta's black lava shoreline and raced to see the find. 'It lacked the antediluvian look of Günther's old engravings', he remembers. 'But when the animal raised its head three feet in the air, I recognized the primeval stare of the British Museum specimens.' Sure enough, it was a big tortoise and a male.

Hamann and Pritchard carried it out into the open to take photographs in a more natural setting. The tortoise didn't warm to this impromptu shoot and stomped for cover. That evening Calapucha cooked up a fine meal of spit-roasted goat. Sitting out together in the open air, the new arrivals tucked in, musing over the tortoise tied up a stone's throw from the camp.

Figure 1.3 Lonesome George moments before leaving Pinta

The following morning, the goat work being finished, the park wardens struck their tents and began to load equipment onto Hamann's boat. One of the last things to be carried from the island was the giant tortoise. The wardens untied the rope around his leg, turned him upside down and carried him across the beach to a dinghy bobbing in the shallows. Cruz sat with the reptile as he was rowed out to the waiting vessel. Hamann and Pritchard stood in silence on the beach, looking on as the tortoise left Pinta and moved slowly out to sea.

Figure 1.4 Lonesome George arrives at the CDRS on Santa Cruz

Back at the CDRS, he was unloaded and carried inland. A routine once-over returned a clean bill of health. Next a thorough inspection and wash prevented the inadvertent introduction of seeds or parasites from Pinta that might be trapped in a fold of flesh. For example, tortoises carry ticks. Just as tortoises on one island are treated as distinct from tortoises on another island, so too are their ticks.

It was down to CDRS director Peter Kramer to decide where to put this welcome guest. The Charles Darwin Foundation, which oversees the research station, had come into being in 1959 to coincide with the 100th anniversary of the publication of Darwin's *On the Origin of Species by Means of Natural Selection*; its tortoise-rearing facilities were still fairly limited. The early years were spent collecting basic data to establish some conservation priorities. Surveys throughout the archipelago revealed that only three tortoise populations could be left to their own devices. The others needed help.

Initially, the research station focused on collecting eggs from Pinzón, where black rats had taken over and were eating baby tortoises. The researchers hatched these eggs, reared the Pinzón babies in captivity and returned them to the wild when old enough to survive the ratty onslaught.

Then during the 1960s, the CDRS took in a handful of individuals from Española. Small enclosures built from blocks of lava prevented these tortoises from escaping and kept them apart. In January 1970, the San Diego Zoological Society funded a new building at the research station. Here tourists could get a close-up glimpse of baby tortoises being reared at the CDRS.

Kramer decided the Pinta tortoise needed to be kept separate from other tortoises at the research station. He gave the honoured arrival free run of a large enclosure down by the sea.

Back on Pinta, just a few days after the tortoise left for Santa Cruz, Peter Pritchard stood on the highest point of the island, his young son in a hip-sling, and stared out to sea. His assistant gazed down the volcano's bright green slopes. Sea turtles were now playing second fiddle to giant tortoises.

Pritchard and his assistant spent their first two days on Pinta scouring beaches for tell-tale signs of turtle nesting.

Between January and March every year, hundreds of green turtles arrive in the Galápagos. Females lumber up out of the sea and into the soft sandy dunes to dig a hollow and lay their eggs, before pulling themselves back into the surf. On one beach, an old track seemed to lead up from the ocean to a nest pit, but the markings were too worn to be sure it had been made by a turtle. There was certainly no evidence of large-scale nesting.

In the absence of any concrete signs of turtles, Pritchard turned his attention to tortoises, specifically the small one that Vagvolgyi had seen four months earlier. On the three-hour trek up the volcano from their camp, they saw no evidence of tortoise life. But that afternoon, taking a different route down, they found some tortoise remains – first a skull, then neck bones and finally fragments of a shell in a deep ravine. These, most likely, were all that remained of a large male. Then a bit further on, they found something even more dismaying – the intact, upturned shell of a smaller tortoise that had clearly been slaughtered. Its plastron or underbelly had been hacked off, presumably so that someone could get at the succulent tortoise flesh inside.

Some years later, Pritchard wrote:

The utter senselessness of it almost reduced me to tears. We were witness to the final extinction of a dramatic and wonderful form of animal life. Instead of poaching one of the more numerous tortoises on a more accessible island like Santa Cruz – or better still, obtaining his meat from one of the goats that settlers were encouraged to kill, a perverse and evil man had made the long journey to Abingdon [Pinta], climbed the mountain, and finding a single tortoise, had turned it over and hacked it apart with a machete.

Over the next few days, Pritchard continued to search for signs of Vagvolgyi's tortoise, without success. He even had two more days than he'd planned for. The boat due to collect them was nowhere to be seen six days on and drinking water was getting perilously low. At the end of a memorable week, the vessel eventually appeared on the horizon. Pritchard had realized his dream of seeing a giant tortoise on Pinta. It was 30 years before he set foot on the island again.

◗

Pritchard held onto the hope that a second, smaller tortoise was still at large. Until about six months later, that is, by which time he was back in Florida. An envelope from Vagvolgyi dropped onto his doormat.

Pritchard pushed his thumb through the seal and drew out a letter from his friend. He unfolded a single sheet of crisp paper. There was the photograph of the tortoise Vagvolgyi had snapped on Pinta in December 1971 and described to Pritchard over dinner in his cramped cottage in Puerto Ayora.

Vagvolgyi's specimen was not small at all. It was large and it was male. In fact, its peculiar hunchbacked shell and the way in which the plates fitted across its back looked just like those of the tortoise now at the CDRS.

The following year, Pritchard returned to the Galápagos at the start of the turtle-nesting season, and took Vagvolgyi's photo with him. One of his first ports of call was at the enclosure of the Pinta tortoise. CDRS director Peter Kramer stood beside him as they looked up from the photo to the tortoise and back again to the photo. Vagvolgyi's tortoise and the animal found in April 1972 were indeed one and the same. The creature that stood before them happily munching on a tasty frond could, they realized, be the last Pinta tortoise on the planet.

Word of the discovery had already reached the outside world. It was not long before he had a name. In the early 1950s, the young comedian George Gobel had taken the

United States by storm with his widely acclaimed television show *The George Gobel Show*, which ran from 1954 until 1960. 'George is about as short and inconspicuous-looking as it is possible to be without being Mickey Rooney', wrote one TV critic of the fresh-faced new talent. 'His speaking voice is as impressive as that of a bored grocery clerk. And he has all the get-up-and-go of a homeward-bound commuter.'

In his show, Gobel frequently acted the part of a hapless, hen-pecked husband with a natty little catchphrase: 'Well, I'll be a dirty bird.' In time, this endearing character and even Gobel himself became known by a TV-hungry American public as 'Lonesome George'. The similarities between Gobel's on- and off-screen personae and the moving story of the solitary Pinta tortoise were obvious.

Just before Gobel bowed out of the show business spotlight for good, the American media began to refer to the Pinta tortoise as Lonesome George. It stuck.

◖

Nobody can be certain of Lonesome George's age. Giant tortoises live longer than anyone has kept records, so we're not even sure how long they might live. The few captives with a fairly reliable history suggest that giant tortoises could be the longest-lived animals on earth.

Some believe that you can tell the age of a tortoise simply by counting up the number of rings on the plates of its shell just as you would tick off a tree's annular rings. Unfortunately, things aren't that simple. Counting rings seems to work quite well for youngsters, but only up to the age of about ten. After that, it is too unreliable to be of much use.

The best we can do is make an educated guess at Lonesome George's age. When he was discovered on Pinta in 1972, he was already pretty large. It takes a giant tortoise somewhere between 20 and 30 years to reach adult size, so this means that he is at least 50. When he was found, the rings on his shell were fresh and light – a sign that he had not yet reached a great age. So he's almost certainly between 50 and 200 years old and most probably somewhere in his eighties.

The most reliable candidate for the title of 'the oldest known tortoise' is an Indian Ocean giant collected in 1766 from Rodrigues in the Seychelles by French explorer Marc-Joseph Marion du Fresne. Left to the French governor of Mauritius, the creature became known as 'Marion's tortoise'. It died in 1918, making it at least 152 years old.

Other venerable tortoises are more suspect. For example, *Guinness World Records* claim that the oldest was a Madagascar radiated tortoise called Tui Malilia that the English navigator and explorer James Cook presented to the Queen of Tonga in either 1773 or 1777. It's not known how old the animal was when Cook gave it to her, but it is thought to have died in 1966. If true (and it's a big 'if') it would mean that this royal tortoise was at least 189.

Finally, there is a Galápagos giant tortoise called Harriet at the Australia Zoo on Queensland's Sunshine Coast. In the 1990s, Harriet's custodians made the startling claim that she had been one of four baby tortoises collected in 1835 by Charles Darwin and his colleagues on HMS *Beagle*. It was such a wonderful story that most people believed it. But in 2004, detective work by journalist Paul Chambers strongly suggested that this was wishful thinking.

Chambers tracked down the results of a DNA test that had been conducted on Harriet in 1998 by geneticist Scott Davis at Texas A&M University. This showed that she had almost certainly come from Santa Cruz, an island that was not on the *Beagle*'s itinerary. Further probing revealed some fairly significant flaws in the story of how Harriet was supposed to have ended up in Australia. Nevertheless, it still looks as though this tortoise has reached a ripe old age. 'Harriet's DNA shows significant differences to that of tortoises living today on Santa Cruz, suggesting that she was born before the great cull of this species that took place in the years after Darwin's visit', wrote Chambers in *New Scientist*. She may be at least 170.

Whatever George's exact age, it looks as though he's a youngster by comparison. When he arrived at the CDRS, he was in pretty good physical condition. This hasn't always been the case.

In around 1980, George suffered a fall and word spread that he had died and the Pinta tortoise had become extinct. He was, in fact, hurt quite badly but with a bit of tlc, pulled through. Soon after his arrival, he also became worryingly fat, allegedly owing to an overindulgent warden; then Lonesome George developed a swelling in his neck, possibly because of a hormone imbalance; and on a couple of occasions a cactus has toppled over in his enclosure, he's eaten the entire thing, suffered from a nasty bout of constipation and gone off his food. Vets have been able to bring him through most of these scares and in the 1990s a nutritionist put him on a strict diet boosted by mineral supplements that helped him lose weight.

In short, George's health is mostly tip-top. For a chap in his eighties, he really should be in his sexual prime. Yet something about Lonesome George is not quite right.

Chapter 2
LONESOME GEORGE'S GIRLFRIEND

Until the 1990s, Lonesome George spent most of his captive life isolated from all other tortoises at the Charles Darwin Research Station (CDRS), alone in his spacious pen on the coast. As the Galápagos tourist industry and George's popularity grew, he needed a more prominent compound. In 1988, Linda Cayot, head of reptiles at the CDRS, floated designs for a new enclosure at the end of the tourist trail through the research station. Work was completed in early 1992 and George took up residency. At the same time, Cayot took a big step. She put two females into George's new pen – females from the northernmost volcano on Isabela Island.

Crossing a male of one population and a female of another is usually taboo in conservation circles. The concern is that it will result in 'outbreeding depression', with unhealthy or infertile young bringing breeding to a standstill. This might arise because each population is adapted to slightly different environments and hybrids can be unfit to survive in either.

In the middle of the 20th century, conservationists in Czechoslovakia decided to let Austrian ibex loose in the Tatra mountains to replace the decimated local population. A little later, they supplemented the Austrian animals with a few ibex from the deserts of Turkey and Sinai. The hybrids that resulted were a muddled bunch. They rutted in the autumn and dropped kids in February – the coldest time of the year – three months earlier than the native ibex. With few offspring surviving, the hybrid population didn't last.

That said, efforts to avoid hybridization can be misplaced. In 1980, the US Fish and Wildlife Service brought 17 red wolves into captivity from the wilds of Texas and Louisiana. These, they believed, were the last in existence, perilously close to extinction because of a combination of habitat destruction, persecution and hybridization with the common coyote. Offspring from a successful (very expensive) breeding programme were returned to some 500 km^2 of northeastern Carolina. The Fish and Wildlife Service monitored the movement of these precious animals closely and intervened to prevent further mixing with coyotes.

Figure 2.1 Conservationists went to great lengths to prevent red wolves hybridizing with coyotes

Then in 1996, a geneticist dropped a bombshell. Close inspection suggested that the red wolf was a hybrid itself, a mixture of gray wolf and coyote. Had all that time and money spent trying to stop red wolves hooking up with coyotes been completely wasted? In 2000, a more thorough genetic screening changed the picture again. The red wolf was, in fact, distinct from both the gray wolf and coyote. It was almost genetically identical to the relatively abundant Algonquin wolf of Canada, suggesting it had never been particularly endangered. The jury is still out on what these so-called red wolves actually are and how they should be managed.

Figure 2.2 Florida panthers have suffered badly from inbreeding

The story of the Florida panther has a more comfortable resolution. In this case, things were really desperate. The population had dropped to around 30 animals and it looked like inbreeding was causing all sorts of health and fertility problems in the few offspring being born. So in 1995, the biologists in charge decided to embrace hybridization, hoping that a bit of gentle outbreeding would inject some much-needed genetic variety into the population. They released eight females from what they believed to be a closely related population in Texas. This intervention was controversial: some felt that the genetic integrity of the Florida panthers was being breached. Fortunately breeding restarted almost immediately, paving the road to recovery and a consensus has emerged that this was a compromise worth taking. Ironically, a recent genetic comparison of Florida panthers and the Texas cats shows they are effectively the same species. It wasn't a controversial introduction at all, just good management.

Initial attempts at captive breeding giant tortoises didn't pay much attention to where animals were from, so inadvertently encouraged hybridization between island types. In 1928, the New York Zoological Society sent Charles Haskins Townsend, director of the New York Aquarium, to the Pacific in a last-ditch effort to save the Galápagos tortoises from extinction. Townsend brought back 180 individuals from several islands, distributing them between zoos across the US in the hope that some would breed in captivity. Not many did. San Diego Zoo received the most animals, but until the 1970s only around 10% of eggs were fertile and only 7% of them hatched. This poor record was almost certainly because captive breeding is hard, especially when animals are outside their native environment. It's also possible that things didn't work because hybridization between tortoises from different islands simply isn't efficient.

Early efforts to breed tortoises at the CDRS faced similar difficulties. According to a 1976 edition of *Noticias de Galápagos*, eggs from the only mixed breeding herd at the research station were mostly infertile. What's more, about a third of the young that did hatch were albinos and short-lived.

So if Lonesome George and one of the Isabela females did get it together, their offspring might suffer from being hybrids, or they might not. Since his options were limited to say the least, experimenting with the Isabela females was worth a go. Unfortunately, he snubbed them. Diligently. He crept around his pen, carefully avoiding contact. Any confrontations that did occur were aggressive rather than amorous. There was certainly no sign that George intended to mate.

The lack of sexual activity in George's corral has led to all sorts of speculation. 'Visit Lonesome George, the world's oldest living gay turtle', touts one American travel agent in the blurb about its Galápagos cruise. Homosexuality could indeed explain why he has refused to court the Isabela females. But scientists are, on the whole, a conservative bunch. They don't talk up homosexual behaviour in the animal kingdom, primarily because, from an evolutionary standpoint, it doesn't make obvious sense. Heterosexuals are more likely to leave descendents than homosexuals.

This kind of argument ignores the fact, so ably explored by Bruce Bagemihl in his book *Animal Exuberance*, that homosexual behaviour is common in the animal kingdom, from the enthused humping of male bighorn sheep to the famous clitoris-rubbing bonobos. If biologists ever attempt an explanation of such behaviour, they tend to cast it in language they can understand. The purpose of such same-sex encounters, they argue, might be a battle for dominance, a contest of stamina, a barter for food, a show of aggression, appeasement, confusion, relaxation, play, greeting. In fact, it could almost be anything as long as it's not plain and simple homosexuality. Might Lonesome George be gay? Since he's spent most of his captive life alone or housed with females, it's hard to tell.

Another possibility is that he's seen so few other tortoises that he isn't really sure how to interact. If a friendly chat about the birds and the bees is helpful for humans, then maybe tortoises need some kind of equivalent – the finches and the frigatebirds, perhaps. The Isabela tortoises may well have been the first females that Lonesome George ever set eyes on (more of which later).

There is a growing awareness that an animal's formative years can have a dramatic effect upon sexual behaviour when adult. This is nicely demonstrated by fostering experiments in birds, where eggs or very young nestlings are taken from one nest and placed in another, sometimes of a different species. Provided the parents accept this intrusion, they will foster the imposters until big enough to go it alone. When these chicks come to breed themselves, they show a preference for mates that look like their foster-parents rather than their genetic parents. Clearly, early learning can shape the way animals think about and approach the opposite sex. If this is the case, Lonesome George may have missed out on some important lessons.

Another related possibility is that some form of male competition is needed for a male tortoise to get into A1 reproductive condition. In their efforts to establish dominance (and so possibly attract mates), males go in for a very characteristic ritual combat. Face to face, each raises his head as high as he can, occasionally delivering a downward bite to his opponent's head. The fight comes to an abrupt end when one tortoise proves he has higher reach; the loser then usually makes off at high speed.

Perhaps watching other tortoises courting and mating might give George some ideas. In animal-breeding circles, it is quite respectable to give both males and females the opportunity to engage in a bit of voyeurism. For example, a bull will be keen to mount a female and will ejaculate more quickly if he has just seen another pair of animals at it. He will even respond in this way if he is just in the presence of another male. This kind of experience probably changes how a male perceives the female before him. If he senses he could lose paternity to another, he gets tetchy and overwhelmed by the need to release large numbers of sperm. Farmers exploit this phenomenon to make their male animals faster and more frequent ejaculators.

There has been less work on female animals' response to watching sex. This is probably, in part, because of a (human) male-biased research agenda, but also because measuring the female responses to such stimuli is a lot harder than counting sperm. This doesn't mean that the responses aren't there, as a neat experiment on Japanese quails illustrates. Females penned up with two males tend to show a preference for one. If they then watch a video of the other male copulating with another female and are given a choice between the two males again, they prefer the one they watched on TV. It is not at all clear why a female might change her mind in this way, but it is fair to conclude that female quails are affected by watching sex and seem to prefer males that have demonstrated their sexual dexterity. Or it could just indicate that humans are not the only animals with a weakness for TV celebrities.

Thomas Fritts, a US zoologist and one-time president of the Charles Darwin Foundation, has spent a large chunk of his life thinking about Galápagos tortoises. In 2002, he wrote a detailed report for the CDRS on the opportunities for the recovery of the Pinta tortoise in which he recommended that George interact with other males. 'Some animal species show increased interest and potential for mating', he wrote, 'when stimulated by contact and competitive interactions with others of the same sex.' Fritts suggested that another male should be put into Lonesome George's enclosure for one or two hours a day 'with constant visual monitoring to prevent excessive interactions and to ensure that no mounting of females by the extra male occurred'.

Alternatively, a new enclosure could be constructed adjacent to Lonesome George's corral. 'Potentially the visual, olfactory, and even auditory stimuli that might be experienced due to the close proximity of the pens could elevate the reproductive interest of the Pinta male without risk of either unintended insemination of the prospective mates or risk of injury due to direct contact with other males', Fritts urged.

So exposing Lonesome George to some sexual imagery, either on screen or in the leathery flesh, might be no bad thing. Nobody has tried it, though. Perhaps because so many tourists pass through the CDRS every day? The star attraction watching tortoise porn on a vast widescreen TV is just not the sort of image the research station wants to project.

●

The only other place you'll find giant tortoises outside a zoo or private collection is on the far side of the world in the Indian Ocean. Efforts to breed tortoises native to the Seychelles archipelago support the idea that social life can be crucial. In the late 1990s, Justin Gerlach, the scientific coordinator of the Nature Protection Trust of the Seychelles, spotted six Seychelles tortoises and six Arnold's tortoises mixed in with captive populations of the abundant Aldabra tortoise. At that time, these 12 creatures were the only known survivors of two species found nowhere else on earth. By captive breeding, Gerlach hopes to restore these animals to the small island of Silhouette. This could be the start of a larger reintroduction effort that might one day see the Seychelles and Arnold's tortoises roaming across other Indian Ocean islands once more.

When Gerlach set up his captive breeding initiative, he kept the two sets of six tortoises in separate enclosures. They are, after all, different species, so it made sense to mate like with like. Everything seemed to go swimmingly. The tortoises mated and the females produced eggs. But of some 260 that were laid, not one was fertile.

Then Gerlach spoke to Owen Griffiths of the Vanilla Crocodile and Tortoise Park on nearby Mauritius. Griffiths is one of the most successful breeders of captive giant tortoises in the world. He told Gerlach that a critical mass of 12 animals is needed for success. Fewer and no matter how many times the reptiles mate or how many eggs they lay, none is fertile. With nothing to lose (except perhaps genetic purity), Gerlach lumped his tortoises together to make the magic dozen. Despite their different evolutionary origins, this appeared to do the trick: the very next clutch laid was fertile. In October 2002, the first hatchlings were born. Some of these are probably hybrids, but others could well be pure Seychelles or Arnold's tortoises.

Not even tortoise breeder Griffiths knows if there's anything to his hunch, but perhaps an element of voyeurism for both sexes and sparring between males is essential. It certainly seems to have worked for Gerlach and his efforts to preserve and reintroduce the Seychelles and Arnold's tortoises to Silhouette and beyond. He already has more than 100 hatchlings.

Figure 2.3 Three-month-old Seychelles giant tortoise from the captive breeding programme on Silhouette

Although the CDRS has never explored such indirect strategies to coax Lonesome George into reproductive action, it has sanctioned a more direct approach. In early 1993, only months after the Isabela females joined Lonesome George and while disappointment at his inactivity was still fresh, Gisela von Hegel, a German zoo vet, visited the Galápagos to help out on a long-term study of marine iguanas. While on a 10-day field trip, she got chatting to Linda Cayot, who had just been promoted to head of protection at the CDRS. The vet planted the seed of an idea in Cayot's mind that would flourish and grow months later in the hands of young Swiss graduate Sveva Grigioni.

She described and demonstrated a technique, common in the zoo world, used to examine a male tortoise's 'intromittent organ' – his penis. In Chelonia (tortoises, turtles and terrapins), the penis is kept out of sight inside the tail when it's not seeing action. This helps male turtles to swim in straight lines and male tortoises to clamber across rough terrain without undue discomfort. Indeed, an inappropriate erection can lead to severe injury that may leave a male unable (or unwilling) to attempt copulation. When aroused, the penis – which at this stage is more like a strip of muscle – emerges from the cloaca (a reptile's dual-purpose, lower-end orifice). Upon erection, the organ curls along its length to form a tube down which semen runs.

Growing up on rocky Pinta, Lonesome George could have sustained some kind of injury. The penis-exposing trick, von Hegel suggested, would be worth trying on him just to check. It might also raise his sexual awareness and yield some sperm. In an enclosure containing tortoises of unknown origin, von Hegel selected an animal on which to demonstrate. She began to touch his rear end and stroke his legs, causing the beast to raise himself off the ground. Then she began to caress his tail. Eventually the penis flopped out and with more gentle rubbing produced an ejaculate. 'She could do this in just a few minutes', says Cayot. Cayot made a mental note of the technique, but with the CDRS understaffed and other projects jostling for attention, it was hardly a priority.

—

Seedy as it sounds, semen collection from animals is remarkably common. Techniques are most advanced in the livestock industry, where commercial breeders commonly collect sperm for artificial insemination. This gives them control over the timing of reproduction and a say over which genes the next generation will carry.

There are many variations on the theme, and they usually involve the manipulation of male genitals. The exception to this rule is to allow a natural copulation and then take a spoon (or similar) to the vagina (or similar) and scoop out the semen. This low-tech method, sometimes used for livestock, is pretty reliable and can yield large numbers of sperm. But cells in an ejaculate recovered in this way are often sluggish because of the alkaline conditions in the female reproductive tract.

A better technique, as far as ejaculate quality is concerned, requires fitting the female with a condom of sorts, into which the male dispenses the goods. This gives a substantial ejaculate and because the sperm are protected from the female tract, they are still pretty nippy. It does have its drawbacks. Devices can turn inside out when the male withdraws if they're not fitted properly. To avoid such problems, an artificial vagina can be slapped onto the male's penis before he enters the female, although in general, the quality of the ejaculate does not match that obtained from a real vagina.

Then there are methods that dispense with the female altogether. Manual stimulation (of animal by human) is pretty standard, although conducted with extreme care if the animal is a bull. In this scenario, a dummy female is often used.

This is usually rather a basic but sturdy structure that allows the handler some protection from the animal, but it takes time to train an animal to mount a false female.

For poultry, the methodology is a little bit different because, unlike mammals, most birds do not have a penis. Sex for our feathered friends is more like a genital kiss; male and female bring their cloacae together for a moment (usually no more than a second) while sperm are transferred in an urgent flurry of wings. With nothing to hold onto, chicken and turkey farmers must give males a well-practised abdominal squeeze.

Conservation scientists use similar techniques, often to propagate poorly represented genes. For many wild or captive animals, things are pretty challenging; some species are more reluctant than others to let someone manhandle their manhood. In many cases, manual stimulation or any of its many incarnations may not be a realistic option and electroejaculation (which we will come to) is the method of choice. But while on the subject of manual stimulation of exotic species, it is worth recounting the experiences of Thomas Hildebrandt of the Institute for Zoo and Wildlife Research in Berlin.

There cannot be many people prepared to collect an ejaculate from a bull elephant by hand. Hildebrandt is one. He heads up one of two teams that have collaborated to advance artificial insemination for elephants at a terrific pace. The animals are not endangered – their health is.

Several years ago, it was discovered that if young adult female elephants do not get pregnant quickly, they develop leiomyomas. Although benign, these uterine tumours can grow to weigh as much as a woman, and as Hildebrandt puts it, 'look like the greatest Swiss cheese'. Whilst they don't kill elephants, leiomyomas give them a lot of pain, vaginal discharge and increase their susceptibility to infectious disease. So if elephants are being kept in zoos (and, like it or not, they are), something must be done to get females pregnant, and quickly.

Only a few bull elephants are held in captivity. Males are much more dangerous than females – so most captive elephants are female. Transporting the occasional male around the world for each liaison is not practical. Apart from the stress to the animal and handler, it's costly. It's safer and cheaper to cart a cool-flask of semen from A to B than to ship a seven-tonne animal with an attitude. What's more, moving animals around runs a major risk of transmitting infectious diseases from zoo to zoo. In the US, 10% of captive elephants have contracted tuberculosis as a direct result of transport and 30–35% of baby elephants die of herpes. So while there are male and female elephants at Jerusalem and Tel Aviv zoos, they can never meet because the Jerusalem animals have a severe infectious eye disease and the Tel Aviv animals don't. A courier service to take sperm from one group to the other is the only way they are going to reproduce.

Hildebrandt has perfected a manual technique to get at and stimulate a captive male's internal glands by way of its rectum. Incredibly, this can be done without anaesthetizing the elephant, and perhaps because of the pleasurable sensations associated with the massage, bull elephants will tolerate it really quite well – far better than someone trying to take a blood sample, for example. So well, in fact, that this can be performed up to five times a day. Incidentally, Hildebrandt offers a stark warning: 'the penis is actually a structure that you never should touch'. Any contact will elicit an eager search for the vagina, he says. 'This searching movement is so powerful that a colleague of ours got a black eye and somebody else got hit too. Touching a penis is not helpful.'

So Sveva Grigioni was in good company. She began by practising on tortoises less precious than Lonesome George.

'Sveva could get the other male tortoises in the exhibition pens to produce sperm within 10 minutes', recalls Cayot.

When, after several days honing her skills, Grigioni entered Lonesome George's enclosure, things did not move so fast. 'He was very shy at the beginning', she explains. 'He is such a big animal and he was so afraid.' So like all good handlers of wild animals, she set about gaining his trust. Routine was important. Grigioni would enter George's enclosure at the same time every day and spend one to two hours with him. At first, she just sat at some distance, but as each day passed she moved nearer – as close as he would allow. Eventually she could stroke his shell without him withdrawing into it. Grigioni had won his trust and her nickname – 'Lonesome George's girlfriend'.

Figure 2.4 Sveva Grigioni at work with Lonesome George

In her patient efforts to free Lonesome George's dormant sexuality, Grigioni used an extra tactic: pheromones. Every day, she washed herself with a neutral soap to reduce her human scent that might be a tortoise turn-off. Then she covered her hands with genital secretions from the Isabela females. These may have contained sex pheromones – sexually charged odours that pass between individuals of the same species.

The classic sex pheromone is bombykol, the devastatingly alluring chemical released by female silkworm moths to entice males from great distances. Males are so sensitive to this pheromone that, as Lewis Thomas put it in *The Lives of a Cell*: 'if a single female moth were to release all the bombykol in her sac in a single spray, all at once, she could theoretically attract a trillion males in the instant'. For vertebrates, these chemical missives usually carry more information than a straight-forward 'come and get me' signal; they often reveal the reproductive condition and genetic suitability of the sender.

A surprising experiment involving smelly T-shirts even suggests that humans produce sex pheromones that could be important in choosing a suitable partner. In 1995, evolutionary biologist Claus Wedekind handed out a small cardboard box to male volunteers. Inside was some perfume-free soap, a T-shirt and a short list of instructions: they were asked to wash thoroughly using the soap, wear the T-shirt non-stop for two days (without washing again), return it to the box and shut the lid – phew! Female volunteers were then asked to sample the odour of six boxes chosen at random and judge which they preferred (or, more probably, disliked the least). Wedekind found that women would consistently select smells from men with different immunity genes from their own. Sniffing out genes in this manner could be a useful way to give children the most robust genetic start possible, he suggests.

Sex pheromones also carry information between individuals of the same sex. Just as visual cues like the presence of competing males can increase a male's urge to get frisky, so too can olfactory cues. A rather startling experiment in sheep demonstrates this quite nicely. Mexican researchers, looking to increase sperm yield from prize rams, found males nearly twice as quick to copulate with a female if her vulva had been smeared with a rival male's semen rather than their own. So although Grigioni only used female secretions to entice him, it could be that pheromones from another male might also rouse the fire in Lonesome George's loins.

Good-humoured Grigioni succeeded where nobody has before or since. First, she confirmed that Lonesome George's reproductive anatomy looks normal. Second, she got him to begin showing signs of sexual activity. 'Day by day, he started to be more interested in the females', she recalls. 'He started to try copulation but it was like he didn't really know how.'

Just then, Grigioni's time at the CDRS came to an end, giving 'coitus interruptus' a whole new meaning. She had to return to Europe to begin her doctorate. She never finished what she started and Lonesome George never mated with either of the females. 'If I had had more time, perhaps I would have had more success', she says. 'I am almost certain that I would have got sperm.'

An eventful four-month chapter in Lonesome George's life came to an abrupt end. When Grigioni left, he slowly returned to his former stubborn ways. Occasionally, he will show some interest in the Isabela females, but nothing like he did during the halcyon days of Grigioni's stay.

Chapter 3

THE ORIGIN OF A SPECIES

Lonesome George's enclosure is a grand affair, clearly designed for the crème de la tortoise crème. He (and his female co-habitees) have free rein over a comfortable area of the research station. There is plenty of vegetation in which George can find the privacy he evidently seeks, tree-like cacti, a shelter which he uses during the hot season and a large pool wherein he is free to bathe (should he wish to venture out into the open). And at some distance, his fans – the tourists – skirt respectfully around. They peer into the undergrowth, their cameras poised should they be lucky enough to catch a glimpse of the world's most famous giant tortoise.

The best days to visit are Mondays, Wednesdays and Fridays. George likes his food and these are the days when he is fed. When the warden enters his enclosure bearing edible gifts, George will march out from behind shrub cover to meet him or her, brimming with hungry confidence. Then, and often only then, it is possible to get a really good look at his impressive 90-kg bulk. With his head withdrawn, he's not much more than a metre from nose to tail. It is when he's feeding, and his neck telescopes out as far as it can go, that he assumes an almost regal grandeur.

Giant tortoises are striking creatures. The first time you see one (whether captive like George or still roaming wild in the highlands of the islands), you cannot help but be impressed, just as the very first visitors were in the early 16th century. The man credited with discovering the archipelago in 1535, the Bishop of

Panama, Fray Tomás de Berlanga, described seeing 'such big tortoises that each could carry a man on top of himself'.

One key question feeds the awe that Galápagos tortoises inspire. How on earth did they get there? During the 19th century, several radically different solutions to this conundrum came and went. Then, towards the end of the century (as we shall see), the weight of opinion began to gather behind the idea that tortoises had floated out from Central or South America and washed up on the remote shores of the Galápagos.

For some, this journey seemed just too incredible. One such was William Beebe, a Brooklyn-born, self-taught naturalist and explorer, who made two expeditions to the Galápagos in the 1920s. Amongst much else, he wanted to know if giant tortoises could have survived the 1000-km journey from the continent. He became one of very few people to have tested the seaworthiness of a giant tortoise.

Figure 3.1 Members of the *Noma* expedition in Darwin Bay on Genovesa. William Beebe rests his hands on a rifle (right)

First, Beebe had to find a likely victim. By the time he visited the Galápagos in 1923, most tortoise populations were shadows of their former glory; on some islands only very few remained. In the four days that Beebe's party spent in the archipelago, they found just one. They were in the crater of Pinzón crawling on hands and knees through thorny scrub when up went a yell. They lashed the unfortunate reptile onto poles and carted it slowly over the terrible lava, through the thick growth of cacti and thorny plants to the shore.

On the beach, Beebe watched as several members of his search party lifted the tortoise into a rowing boat wedged into the sand. Although it was now dark and the temperature had fallen, this was tough work. He removed his cloth hat and mopped the sweat from his balding brow.

One by one, the team clambered aboard. Beebe found himself cramped beside the tortoise, his tall, thin frame bent double. He let a hand rest atop its shell. By moonlight, they rowed out from Pinzón towards the steam yacht – the Noma – anchored a few hundred metres from the shore. The following day, they rowed the tortoise back to land, this time to nearby Santa Cruz. There, they set about filming it performing various tasks.

First, they let it loose on a steep, rough lava slope to see how it coped. 'In spite of frequent slipping it kept obstinately ahead in any direction it had once chosen', Beebe noted in his best-selling book Galápagos: World's End, published the following year.

The team then tried the hapless tortoise on soft sand, flat ground and in various other situations until, it seems, the creature became resigned to its fate. 'After we had handled it a few times it ceased even to hiss or withdraw its head completely under the shell.' But its biggest challenge was still to come.

As the rowing boat drew up alongside the Noma, they heaved the giant tortoise up and over the edge into the water. The tortoise dipped into the ocean, a wave washed over its back. Then its neck emerged to full length, raising its head well above the waves, and the top of its shell bobbed back up and out of the water.

Figure 3.2 The ill-fated Pinzón tortoise keeps its head above water

Beebe was stunned: 'When placed in the water alongside the ship, with a full knot current running, our tortoise floated buoyantly, looked about in various directions, lowered its head well under water and gazed around, then deliberately turned and swam across the current toward the yacht.' When the tortoise failed to get a grip on the sheer side of the ship, it floated with the current towards the gangway. This was no better. 'It then turned and swam against the current toward the rowing boat from which I had launched it', wrote the incredulous naturalist in the *Zoological Society Bulletin*.

Beebe repeated the experiment several times, getting the same result on each occasion. The Pinzón tortoise floated confidently, high above the waterline. 'I could see the throat vibrate in breathing, without any detectable lowering or elevation of the body', he wrote. 'So for a time at least these creatures have perfect control over themselves in the water.'

Might tortoises really have made it from South America to the Galápagos across 1000 km of ocean? This question was soon answered to Beebe's satisfaction: 'A week later this tortoise died without warning', he noted in *Galápagos: World's End*. The body was somewhat wasted, but hardly enough to account for so sudden an end in a slow-living, slow-dying animal. A post-mortem settled the matter. The tortoise had congested lungs, suggesting that it had swallowed too much salt water during its marine ordeal. 'This would negative any possibility of the tortoises being able to make their way over wide expanses of water, either from the mainland or from island to island, in spite of their unusual swimming', Beebe concluded, wrongly as it turned out.

What alternative explanations were there? One possibility mooted in the early 19th century was that sailors had brought giant tortoises to the Galápagos from the only other place they existed – the Seychelles and Mascarene islands of the Indian Ocean. This popular explanation was blown out of the water in 1875 by Albert Günther, director of the Department of Zoology at the British Museum. Günther's careful examination of several Galápagos and Indian Ocean giants revealed clear differences, confirming that the two groups had entirely different origins and were not one species moved around by humans. This was consistent with the emerging consensus that tortoises got to the Galápagos and Indian Ocean islands by floating out from the continents.

Then in 1889, George Baur, a German-born scientist at Yale, put forward a controversial alternative theory. It came to him in the university museum as he unpacked fossils from a recent field trip to Nebraska. Among the specimens were giant tortoise fossils. Lifting one from its box, he stopped still, struck by its resemblance to the giant tortoises in the Galápagos.

Baur began to wonder if the continental and archipelago animals could be one and the same species that had once been able to move freely between America and the

Galápagos. Tortoises walked there, he reasoned, by way of a land bridge, which has since sunk into the sea. 'The Galápagos originated through subsidence of a larger area of land', he wrote in the *Proceedings of the American Antiquarian Society*; 'they do not represent oceanic islands, as generally believed, but are continental islands.'

With our current understanding of geology, Baur's subsidence theory verges on the ridiculous. We now know that the earth's surface is made up of several tessellated plates floating slowly over softer rock beneath. Occasionally, a hot spot under a plate will break through and spew molten rock or magma onto the surface to form a volcano. The smattering of islands and hundreds of rocky outcrops that make up the Galápagos came about because of just such a hot spot. As the plate moved over it, drifting from west to east, so island after island emerged from the sea, lifeless larval outbursts forming the archipelago that we see today.

The oldest volcanoes, which have long since drifted off the hot spot, have been eroded from towering mountains to low and barren islands. Detailed surveys of the Pacific Ocean between the existing archipelago and the South American mainland also reveal a string of drowned islands, so weathered they are covered by waves.

Working more than half a century before the discovery of plate tectonics, Baur did not know this. He and a handful of others stubbornly advocated the subsidence idea, in spite of a wealth of evidence suggesting the Galápagos' volcanic origin. The sink-or-swimability of the tortoise became central in their argument.

Tortoises are unable to swim, pronounced the recalcitrant Baur. Even if one did manage the epic journey, it would never be able to establish a population. This would only be possible if 'a similar accident imported another specimen of *the same species*, of *the other sex*, to *the same island*', he wrote in 1891. 'We would have to invoke thousands of accidents.'

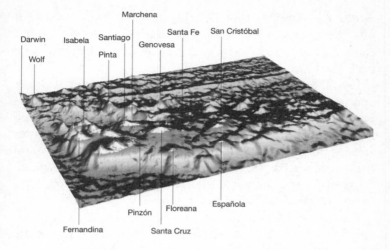

Figure 3.3 3D view of the sea floor topography in the Galápagos. As they drift east, off the hot spot, the islands are weathered down, eventually beneath the waves

Similar objections continued to resurface for decades. John Van Denburgh, the curator of the California Academy of Sciences, was a keen Baur supporter. His thinking on tortoise origins was almost certainly influenced by an extraordinary experience during the California Academy's collecting trip to the archipelago in 1905 and 1906.

On 17 March 1906, four members of the expedition were on Isabela looking for tortoises. They spent much of the day on their hands and knees, crawling through dense scrub near the shore. Eventually they came across three large males, herded two of them down to the shore, floated them out to the skiff and tried to lift them in. The colossal weight and a freak wave caused the boat to capsize. The swell smashed it into a thousand pieces and carried the tortoises out to sea.

The collectors returned to the expedition vessel – the

Academy – empty-handed, exhausted and half-naked. 'All hands are anxious to leave as soon as possible, and I won't shed any tears myself', wrote reptile expert Joseph Slevin once back on board.

The following morning, a surreal sight offered some compensation for the previous day's tribulations. A few pieces of broken skiff drifted past the anchorage, followed by one giant tortoise and then the other. The unfortunate creatures seemed none the worse for their 18 hours at sea. 'They would occasionally stick their heads out of water and look around, but they floated along like a cork, nearly all the carapace out of water', observed Slevin in his journal.

This incident and others like it led Van Denburgh to conclude that tortoises don't swim but are at the mercy of the winds and currents. 'When they drift on island shores, they are usually so battered and injured on the rocks that they only live a few days thereafter', he asserted in a 1914 paper.

Van Denburgh fell back on Baur's subsidence theory, writing:

> We must rather adopt the view that the islands are but the remains of a larger land-mass which formerly occupied this region, and was inhabited by tortoises, probably of but one race; that the gradual partial submersion of this land separated its higher portions into various islands; and that the resulting isolations of the tortoises upon these islands has permitted their differentiation into distinct races or species.

By then though, this was a minority position. The geological evidence for a volcanic origin of the archipelago was stacking up. Explanations of how flora and fauna might reach such lifeless laval outbursts quickly followed.

Much of the early thinking focused on plants. In 1830, the British geologist Charles Lyell proposed that small, matted islands of fallen trees, floating out at sea, could carry all manner of plant and animal life to all sorts of remote places. These rafts, he suggested, 'may arrive, after a passage of several weeks, at the bay of an island, into which its plants and animals may be poured out as from an ark, and thus a colony of several hundred new species may at once be naturalized'.

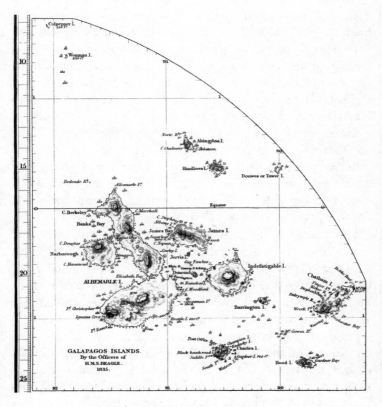

Figure 3.4 Fitzroy's map of the Galápagos

Indeed, as early as 1798, James Colnett, a veteran of Captain Cook's second voyage, had noticed driftwood in the Galápagos much larger than any of the trees growing there. Similarly, in 1835, Robert Fitzroy, the captain of HMS *Beagle*, noted in his account of the expedition that 'driftwood, not the growth of these islands, is frequently found on the south-east shores'. Surely natural dispersal was sufficient to explain the origin of the Galápagos flora and fauna?

Charles Darwin, the naturalist who accompanied Fitzroy on the *Beagle*, certainly thought so. Some time after his return to England in 1836, he set about experimenting with seeds and sea water to work out whether they might survive a long journey at sea. Shut away in his Kent study-cum-laboratory, Darwin weighed out pure sea salts he'd bought from a chemist in Holborn; he dissolved them in a fixed volume of water to keep the concentration of salt constant in his experiments.

By the time he published *On the Origin of Species* in 1859, he had exposed the seeds of 87 different plant species to his sea water concoction. Incredibly, nearly all of them germinated after weeks stewing in brine; some still seemed in perfectly good working order after several months. It certainly looked as though seeds might be able to survive at sea.

There was one snag. Almost all the seeds sank, so would never have floated very far from a continent like South America. Darwin thought a bit more. Seeds are much more buoyant when dry. Plants washed downriver by floods might be cast onto the bank and dry out before being washed into the sea sometime later, he reasoned.

From these simple experiments, Darwin managed to conclude that 'the seeds of 14/100 plants of any country might be floated by sea-currents during 28 days, and would retain their power of germination'. Joseph Hooker, the botanist who Darwin asked to classify his Galápagos plants, lent further support to the idea that the islands' plants had arrived by sea. The species that made it, Hooker hypothesized, should be those with the toughest seeds. Indeed, a high proportion of Darwin's Galápagos plants were just such hardy species.

As well as floating, Darwin elaborated on other means of dispersal suggested by Lyell. All kinds of seeds might be carried out to islands like the Galápagos in earth wedged between the roots of trees. Seeds of other plants might be transported in the intestines of dead animals and yet others could be carried by live birds, either in their crops or perhaps in earth caked onto their bills or feet. Here were four plausible mechanisms of seed dispersal, each backed up by elegant experiments. And there were undoubtedly more, Darwin admitted. Together, such explanations could easily account for the origin of the flora on the Galápagos.

What about its fauna? It is simple to explain how animals with wings reach distant places. Indeed, almost all islands have birds and bats. But working out how earth-bound animals reach isolated islands is not so easy. As Darwin put it: 'there are many and grave difficulties in understanding how several of the inhabitants of the more remote islands … could have reached their present homes'.

Rather than addressing such 'grave difficulties' directly, Darwin did something much smarter. He ignored them. He was more interested in explaining why certain groups of animal were so scarce on remote islands. 'I have not found a single instance, free from doubt, of a terrestrial mammal (excluding domesticated animals kept by the natives) inhabiting an island situated above 300 miles from a continent or great continental island', he observed. If terrestrial mammals had trouble surviving at sea, which seemed possible, this would explain why remote islands are such mammal-free places. Similarly, Darwin reasoned that frogs, toads, newts and other amphibians are absent from volcanic islands because these freshwater creatures can't tolerate salty water, so simply wouldn't have made the long swim.

As usual, Darwin was essentially spot on. That said,

amphibians have occasionally made it out to volcanic islands. The island of Mayotte in the Indian Ocean is part of the Comoros archipelago in the Mozambique Channel between Madagascar and continental Africa. It has two species of frog.

The assumption was that humans brought them to the island. DNA evidence indicates otherwise. The Mayotte frogs are like no other species on earth. Not only are they genetically distinct from those on Madagascar and the African continent, they are distinct from each other. This suggests that frogs reached Mayotte on two separate occasions and, because of the isolation that island living provided, they evolved into the two unique species we see today.

This kind of exception doesn't undermine Darwin's conclusion that remote volcanic islands are typically populated by a handful of hardy plants and a collection of non-mammalian animals. They are colonized, in other words, only by those species that manage to survive a journey over open ocean.

Beebe was one of the last notable Galápagos figures to hold out against this view. The swimming experiment on the ill-fated Pinzón tortoise had made a strong impression. 'I am in perfect agreement with Van Denburgh in regard to the formation of the Galápagos by subsidence', he wrote in *Galápagos: World's End*. 'I go even farther and can see no explanation of the origin of the flora and fauna except through a former, direct connection with the mainland.'

Quick to counter this kind of argument was Charles Townsend, director of the New York Aquarium. On his early expeditions to the Galápagos – in 1888 and 1891 – Townsend took part in the first attempts to sound out the depths of the Pacific. These expeditions found that the ocean was more than 3 km deep in most places between the continent and the Galápagos. This scuppered the idea of a land bridge. In an insightful analysis published in 1925, Townsend reworded Darwin's logic on the islands' spartan flora and fauna. If there had been a land bridge, surely more plants and animals would have reached the islands, he pointed out: 'It is pertinent to inquire, why was it so little used?'

Like Darwin, Townsend was of the opinion that, given enough time, 'it is inconceivable that various forms of living flotsam could have failed to arrive'. If tortoises could, at a pinch, float – which the experience of the California Academy scientists and Beebe confirmed – then the prevailing currents would do the rest. The Humboldt Current, which runs up the coast of South America, diverts westwards at the equator directly to the Galápagos.

Townsend laid out the case for the migration of tortoises on the prevailing westerly currents. 'The ancestry of the island tortoises need not be regarded as mysterious, with so close a relative as *Testudo tabulata* living no farther away than the Isthmus of Panama', he wrote. These mainland tortoises were not giants, but with some recorded at over 60 cm in length, they were not small either. 'The origin of the Galápagos tortoises is directly connected with the species *Testudo tabulata* of northern South America', Townsend concluded.

Such ideas were impossible to test, until the advent of DNA profiling in the mid-1980s, that is. Geneticists at Yale University collaborating with a host of others around the world got on the case in the 1990s. They collected DNA from several tortoise species on mainland South America. They hoped to find out which showed the greatest genetic similarity to Galápagos tortoises to pinpoint roughly where these giants had originated and what their ancestors might have looked like. The red-footed tortoise *T. tabulata* (now known as *Geochelone carbonaria*) was one of the mainland creatures they analysed. Was Townsend correct in supposing that this species and the Galápagos giants were 'directly connected'?

Townsend had the right idea, but DNA profiling indicates

that he was wrong to opt for *T. tabulata*. The giant tortoises of the Galápagos have more in common with the Chaco tortoise *Geochelone chilensis*, which lives in Bolivia, Paraguay and Argentina. The Yale analysis does not show that the Chaco tortoise evolved into the Galápagos giant. This would be like saying that chimpanzees evolved into humans, which they didn't. Instead chimps and humans sit near each other on the primate family tree, like leaves on two neighbouring branches. The Chaco and Galápagos tortoises occupy similar positions in the tortoise family tree. The Chaco and the Galápagos branches have been growing their separate ways for several million years. But way back, at the point where these branches fork, a leaf once grew – the common ancestor of the Chaco and Galápagos lineages.

We don't know what this common ancestor looked like or how it behaved because it no longer exists. But we do know that it probably lived between 6 and 12 million years ago. Perhaps, like today's Chaco tortoise, it roamed across the dry, thorny habitats in the heart of South America, feeding on grasses, succulent plants and cacti, conditions very similar to those experienced by the Galápagos tortoise. Somewhere along the Galápagos branch of the family tree, the reptiles began to increase in size until eventually they became the giants we see today.

There is some debate about whether this evolutionary growth spurt happened on the South American mainland or out in the archipelago. Many are convinced it occurred in the Galápagos: small tortoises made it out to the islands and once there, in the absence of anything to stop them, evolved into giants.

This process, known as gigantism, is common on isolated islands; it left some very large creatures in some very small places, although many of these are now extinct. The Mauritius dodo, for example, weighed a whopping 20 kg and had lost the power of flight. DNA collected from museum specimens suggests that this strange bird evolved from a pigeon-sized ancestor. Similarly, Madagascar's elephant bird, which stood 3 m tall and weighed around 450 kg, was probably descended from a comparatively slight ostrich-like bird. Giant tortoises are often wheeled out as another case of island gigantism.

There's good evidence that animals can become large quite quickly once they've reached the isolation provided by an island. Southern Australia is home to several populations of highly venomous tiger snake. Most are of a fairly standard size. On several small islands just off the coast, however, the tiger snakes are huge, reaching up to 160 cm long and weighing well over 1 kg. Those on Mount Chappell Island, for example, can weigh twice as much their relatives on nearby Tasmania only around 50 km away. What's particularly startling is that Mount Chappell Island is less than 10,000 years old, giving very little time for such a radical change in body size.

The extinct Haast's eagle of New Zealand, with its incredible wingspan of up to 3 m, is another illustration of how quickly gigantism can evolve. Analysis of DNA from bones revealed that it was most closely related to some of the world's smallest eagles. This suggests that the ancestors that first colonized New Zealand were regular-sized raptors, but in a relatively short time had evolved into the awesome Haast's eagle.

What is there to stop tortoises undergoing a similarly rapid change in body size? Not a lot, it seems. They usually reach their maximum size by the age of 30. But Goliath, a Galápagos tortoise who lived out his days at the Life Fellowship Bird Sanctuary in western Florida, just grew and grew. He left the Galápagos as a hatchling in 1961 and passed away in January 2002. In just over 40 years, he had already reached a colossal 422 kg – that's nearly five times the size of Lonesome George – and was still piling on the pounds. Perhaps the molecular machinery that tells a tortoise when to stop growing is rather rudimentary. A few minor mutations of one or two genes might be enough to cause a startling change in body size.

Figure 3.5 A giant Haast's eagle attacks a New Zealand moa

So it may not take much for a small tortoise to go giant. But did this occur on the islands, as most assume? Not necessarily. The fossil record reveals that there were giant tortoises living on the continent until about a million years ago. These hefty creatures would have stood a much better chance of surviving the perilous journey to the archipelago than would regular-sized animals: giant tortoises have good fat reserves, can extend their necks above waves and are better equipped for life in a hot, dry, boulder-strewn world.

Further support for the floating-giant theory comes from Aldabra in the Indian Ocean. Geological evidence shows that this remote atoll was completely submerged around 120,000 years ago. It only re-emerged about 100,000 years ago, yet today is covered with giant tortoises. It is unlikely that small tortoises could have gone large in such a short time.

With everything we know about the geology of the Galápagos, tortoise DNA and the swimming skills of these reptiles, the origins of the islands' tortoises probably lie in a small and hardy ancestor living somewhere in central South America about 10 million years ago. Several million years later, the first islands in the archipelago began to emerge. The mainland tortoises were by now of immense proportions. About 3 million years ago, some of these giants floated out, washed up on one of the easternmost islands and survived.

There are, of course, other plausible scenarios. One is that the drowned islands between the mainland and the present-day Galápagos Islands acted as stepping stones, breaking up the journey into shorter, more plausible hops. There is certainly good evidence that the archipelago's iguanas used them.

There are two main types of iguana in the Galápagos – land and marine – and there are two possible explanations for their origin. Either iguanas reached the archipelago on two separate occasions or they made it out to the islands only once and subsequently evolved into the land and marine forms that we see today.

Once more, DNA differences shed some light on the matter. The land and marine iguanas are more closely related to each other than either is to any mainland iguanas. This suggests that the fork in the iguana family tree took place on an island and not on the South American continent. However, the DNA dates this fork at between 10 and 20 million years ago, long before any of the existing Galápagos Islands had erupted. It's always possible that there's a problem with the dating method, but if there isn't, then these two conclusions are difficult to reconcile. Unless, that is, the fork between the land and marine branches of the iguana family tree took place on intermediate islands that have long since weathered away. If iguanas used such island stepping stones, why not giant tortoises?

Whatever route George's distant ancestors took to the Galápagos, they surely made the death-defying crossing only rarely. Perhaps there was just one trip: perhaps a single female weighed down with eggs lumbered up the beach, probably on San Cristóbal, and laid her clutch in the sand. The rest is evolutionary history.

Chapter 4
RANDOM DRIFT

By the time Charles Darwin came to publish his account of the *Beagle* voyage in 1839 – his *Journal of Researches* – he'd had an important revelation. 'There is every reason for believing', he wrote, 'that several of the islands possess their own peculiar varieties or species of tortoise.'

More than 150 years later, DNA profiling is probing the evolutionary relationships between these different varieties. This has produced some commonsensical results plus a few surprises. One geneticist even believes that Lonesome George's genes raise the worrying possibility that he isn't a Pinta tortoise at all, but an impostor from another island.

There has been a lot of (often very heated) debate over the similarities and differences between the tortoises from different islands and volcanoes and whether they constitute just one or several distinct species. This kind of academic wrangling doesn't really affect those who manage the plants and animals of the Galápagos. Even if everyone agreed that the tortoises from the different islands were just one species, conservation biologists would still do their best to preserve the genetic integrity of each population. In certain situations, however, the survival of a species can be influenced by its name.

Biologists name the animals and plants around them to bring some sense of order to the natural world. This classification process is called taxonomy: a plant is assigned to the plant kingdom and an animal to the animal kingdom, as if they are taken to different floors of a vast library of life. A fish and a mammal (both being of the animal kingdom) are housed on the same floor, but in different rooms. A cat and a dog are put in the same room, but in different aisles; a lion and a tiger in the same aisle but on different shelves. Only different types of tiger (such as Bengal, Sumatran and Siberian) will be together on the same shelf. At least that's the idea.

Occasionally, important differences between two animals are overlooked and a mistake in classification may occur. It's as if a wayward librarian files two distinct creatures on the same shelf, when each should really have a shelf to itself.

This can be disastrous. In 1877, scientists identified two species of tuatara in New Zealand, a rare lizard-like creature found nowhere else on earth. Unfortunately, this work was overlooked and subsequent classification lumped all tuatara together as a single species. So 20th-century conservationists trying to protect these unique creatures saw just one and not two distinct species. This ill-informed management probably led to the extinction of several important populations. Fortunately one small population of the second and less common tuatara described back in 1877 survives on a group of tiny islands to the east of New Zealand. In short, bad taxonomy can kill.

The most commonly used classification of the Galápagos tortoises recognises 14 different types. Three – on the islands Santa Fe, Floreana and Fernandina – are extinct. Eleven types survive, one of which is the Pinta tortoise. If Lonesome George really is the last of his kind, only 10 different types of tortoise will remain once he dies. Five evolved on their own islands – San Cristóbal, Española, Santa Cruz, Pinzón and Santiago. The other five types are all found on Isabela. Here each inhabits its own volcano, isolated not by sea but inhospitable stretches of barren lava.

The classification of the Galápagos tortoises needs to reflect the similarities of these types – tortoises from different islands and volcanoes can produce hybrid offspring. But it also needs to emphasize their differences, most obviously in size and shell shape.

The solution that has become most widely (although by no means unanimously) accepted is to talk about Galápagos tortoises as subspecies of the same species *Geochelone nigra*. This is the biologist's way of saying the populations are almost distinct species but not quite. Given a few more million years evolving in isolation, each of these subspecies should eventually attain full-blown species status: they would be too different to mate successfully. For now, while they are still able to interbreed, the different types are referred to as subspecies. To pursue the library analogy, this is like putting all the Galápagos tortoises on the same shelf, giving each subspecies its own special space. The extinction of the Pinta subspecies, say, would leave a gap on the shelf that could never be closed.

As geneticists probe with ever increasing sophistication, the established tortoise taxonomy is likely to change. For example, there are two main populations of domed tortoises on Santa Cruz, all deemed members of the same subspecies (*Geochelone nigra porteri*). In 2005, the Yale geneticists performed the first detailed inspection of DNA from these two populations. The researchers found differences so great that, in a *Biology Letters* paper, they called for a significant revision of tortoise taxonomy.

Such a revision could influence the management of tortoises on Santa Cruz. One lineage, in particular, is at greater risk of disappearing than the other. The population at Cerro Fatal to the east of Puerto Ayora has suffered considerable poaching recently. Farmers have also whittled away its habitat. An appreciation that this population carries a unique set of genes makes it even more important that conservation biologists do all they can to protect it.

In addition to teasing apart the knotty branches of taxonomy, DNA sequences can shed light on how tortoises came to live on so many different islands and how each population, sometimes separated by just a few miles, came to look and behave differently. In other words, analysis of DNA from animals like the Galápagos tortoises can give a real insight into the very process of evolution itself.

●

Soon after leaving the Galápagos, Darwin knew the islands were telling him something. 'In space and time, we seem to be brought somewhat near to that great fact – that mystery of mysteries – the first appearance of new beings on this earth', he wrote in the 1845 edition of his *Journal of Researches*.

This was code. It was written in such a way as not to offend the majority of his readers who believed that God created the earth and everything on it. But for those seeking an alternative to the Bible story, this passage and others like it were a nod and a wink. Darwin of course had more to say on the origin of species. His book on the subject changed forever the way that we think about the natural world.

Darwin's five-week meander through the Galápagos produced a significant piece of the jigsaw. It's clear from the attention he pays it in his subsequent writing that the archipelago had a strong influence on his thinking. But careful research in 1982 by Frank Sulloway, a historian of science at the University of California at Berkeley, revealed that Galápagos was not, as is often portrayed, Darwin's eureka moment. Far from it.

Profound statements about evolution and the origins of life are virtually absent from the diary Darwin kept in the Galápagos. 'His primary interest was in the geology', says Sulloway. 'There are twice as many pages of geology notes as there are zoology notes.' Fortunately Darwin also realized that naturalists hadn't visited the Galápagos and that this was a great opportunity for collecting. 'He spent a great deal of time trying to make very representative and complete collections of the different branches of natural history', says Sulloway.

That said, Darwin's approach to field study would raise a few eyebrows today. He rode the giant tortoises as if they were horses: 'I frequently got on their backs, and then giving a few raps on the hinder part of their shells, they would rise up and walk away', he wrote in his *Journal of Researches*. He pestered a land iguana busy excavating a burrow on Santiago: 'I watched one for a long time; till half its body was buried; I then walked up and pulled it by the tail.' He gave marine iguanas a hard time too: 'They will sooner allow a person to catch hold of their tails than jump in the water ... I threw one several times as far as I could, into a deep pool left by the retiring tide; but it invariably returned in a direct line to the spot where I stood.' Darwin was also surprised by the tameness of the birds: 'All of them often approached sufficiently near to be killed with a switch, and sometimes, as I myself tried, with a cap or hat. A gun is here almost superfluous; for with the muzzle I pushed a hawk off a branch of a tree.'

In his zoological notebook, written up as the *Beagle* sailed to Tahiti, he dedicated just over four pages to the giant tortoises, making several detailed observations. 'One large one, I found by pacing, walked at the rate of 60 yards in 10 minutes, or 360 in the hour – at this pace, the animal would go four miles in the day & have a short time to rest.' He noted how fast they drink: 'When the Tortoises arrive at the water, quite heedless of spectators they greedily begin to drink: for this purpose they bury their heads to above their eyes in the mud & water & swallow about 10 mouthfulls in the minute.' He was an interested observer of tortoise sex: 'The Males copulate with the female in the manner of a frog – they remain joined for some hours. During this time the Male utters a hoarse roar or bellowing, which can be heard at more than 100 yards distance.' Letting his inquisitive nature run wild, Darwin exercised his real talent for observation.

DARWIN TESTING THE SPEED OF AN ELEPHANT TORTOISE (GALAPAGOS ISLANDS).

Figure 4.1 Charles Darwin pacing alongside
a giant tortoise on Santiago

Some of his rudimentary 'experiments' didn't come to much:
tugging the tail of the land iguana, for example, didn't precipi-
tate any great insights: 'It was greatly astonished, and soon
shuffled up to see what was the matter; and then stared me in
the face, as much as to say, "What made you pull my tail?"'

By contrast, his repeated tossing of the marine iguana into
the sea presented a 'strange anomaly' that required an
explanation. 'I several times caught this same lizard ... and
though possessed of such perfect powers of diving and swimming, nothing would induce it to enter the water.' Perhaps, he
mused, 'this singular piece of apparent stupidity may be
accounted for by the circumstance, that this reptile has no
enemy whatever on shore, whereas at sea it must often fall a
prey to the numerous sharks. Hence, probably, urged by a
fixed and hereditary instinct that the shore is its place of
safety, whatever the emergency may be, it there takes refuge.'

It wasn't a bad hypothesis, but there's a better one. The most likely explanation for this marine iguana's bemusing behaviour is that its body temperature was too low. Cold-blooded creatures like marine iguanas can only afford to swim once they've spent a good while warming up in the sunshine. Nevertheless, Darwin's shark idea reveals his keenness for asking questions and finding answers.

Figure 4.2 Marine iguanas soak up the sun before taking a dip

The seemingly frivolous face-off between Darwin's gun muzzle and the hawk led him to similar thoughts. 'It would appear that the birds of this archipelago, not having as yet learnt that man is a more dangerous animal than the tortoise or the Amblyrhynchus, disregard him, in the same manner as in England shy birds, such as magpies, disregard the cows and horses grazing in our fields.'

Darwin made careful observations on the flora and fauna on the four main islands that he visited: he spent five days on San Cristóbal; four days on Floreana; one day at Tagus Cove on Isabela; and nine days on Santiago. If he did see any live tortoises on either Floreana or Isabela, he made no mention of it. The Floreana tortoise was already on the brink of extinction and although tortoises do come down to Tagus Cove, they were probably not there in the dry season when Darwin passed through. He did, however, see tortoises on San Cristóbal and Santiago.

'In my walk I met two very large Tortoises (circumference of shell about 7 ft.)', he wrote of his first San Cristóbal encounter. 'One was eating a Cactus & then quietly walked away. The other gave a deep & loud hiss & then drew back his head. They were so heavy, I could scarcely lift them off the ground.' Later on, when he was exploring Santiago, he saw lots more. 'Wherever there is water, broad & well beaten roads lead from all sides to it', he noted. 'In the pathway many [tortoises] are travelling to the water & others returning, having drunk their fill. The effect is very comical in seeing these huge creatures with outstretched neck so deliberately pacing onwards.'

Darwin's exposures to the San Cristóbal and Santiago tortoises were separated by a couple of weeks and by his visit to Floreana. There, the vice-governor of the archipelago, one Nicholas Lawson, told Darwin that each island has its own type of tortoise that can be identified from its size and shell shape. Darwin seems not to have paid much attention to Lawson's bold assertion. Indeed, in the first edition of his *Journal of Researches* published four years later, Darwin mentions it only in passing: 'It was confidently asserted, that the tortoises coming from different islands in the archipelago were slightly different in form; and that in certain islands they attained a larger average size than in others; Mr. Lawson maintained that he could at once tell from which island any one was brought.'

Only when the second edition appeared in 1845 – nearly a decade after the *Beagle* voyage – did Darwin's emphasis shift.

He was now prepared to dwell on the significance of Lawson's claim:

> I have not as yet noticed by far the most remarkable feature in the natural history of this archipelago; it is, that the different islands to a considerable extent are inhabited by a different set of beings. My attention was first called to this fact by the Vice-Governor, Mr. Lawson, declaring that the tortoises differed from the different islands, and that he could with certainty tell from which island any one was brought. I did not for some time pay sufficient attention to this statement, and I had already partially mingled together the collections from two of the islands. I never dreamed that islands, about 50 or 60 miles apart, and most of them in sight of each other, formed of precisely the same rocks, placed under a quite similar climate, rising to a nearly equal height, would have been differently tenanted.

Once tortoises had become established on one of the south-easterly islands – probably San Cristóbal or Española – they were then in a good position to branch out to other islands in the archipelago.

No one is sure how they managed it. In *Darwin's Islands, A Natural History of the Galápagos*, published in 1971, Ian Thornton considered the options. It's unlikely that tortoises would wander into the sea and get carried off, simply because they don't normally go down to the shoreline, says Thornton. They do, however, take the odd tumble when negotiating a rocky incline. 'So it is at least possible that tortoises very occasionally could have fallen over low cliffs directly on the sea', he wrote.

CDRS biologist Linda Cayot has another suggestion. In an El Niño year, when the annual rainfall is huge, tortoises are strangely skittish. 'When extremely heavy rain begins to fall, they quickly move to higher ground, whether night or day', she says. Cayot suspects this is an instinctive response to minimize the risk of being caught in fast-flowing rivers that can suddenly appear. Perhaps some tortoises, especially the smaller ones, primarily females living closer to the coast, never made it to higher ground but were washed to sea, she says.

Currents would then have carried them out to sea and with a bit of luck onto the shore of a neighbouring island. After many generations, the original and the new tortoise populations would begin to look and behave differently.

This divergence of populations occurs, in part, because the machinery that copies DNA (to make sperm and eggs) is not infallible. It's a bit like photocopying. Sometimes a photocopier will make three copies when you wanted one, sometimes it will make none and occasionally the copy will come out looking grey, blotched or bleached. Two machines will not produce the same catalogue of errors in the same order over the same time period. If you keep copying the same document on two different machines, eventually there will be a version that differs from the others. Make more copies of this tweaked copy and more differences will accumulate. Similarly, mistakes occur when DNA is replicated.

This type of iterative change is called the 'random drift' of a genetic sequence. Two tortoise populations that start with identical genes but reproduce in isolation will eventually accumulate genetic mutations; the populations will drift apart and may begin to look and behave differently.

There are also non-random genetic changes over time, usually because certain genetic combinations are, by chance, more or less suited to a time or place than others. Although Darwin did not immediately realize it, subtle differences between islands can winnow particular characteristics. The larger islands that rise up to mountainous peaks have lush highlands – their tortoises have the default dome-shaped shell which is fine for an easy-going grazing lifestyle. The smaller, low-lying islands are dryer and the vegetation more bushy. Here, where there is little to browse on, tortoises have a furled lip at the front of their shells and long necks, enabling them to reach up and nibble at shrubs or cactus pads.

The argument goes that on these small and flattish islands, tortoises that could reach up for food fared better than those that couldn't. To use the jargon, long-necked tortoises had a 'selective advantage' over stubby-necked ones. They could eat more, were less likely to die if the going got tough, so stood more chance of having babies. And after many many generations, long-necked tortoises with saddle-shaped shells came to dominate.

Figure 4.3 Domed tortoise from Darwin volcano on Isabela (left) and saddleback from Pinzón (right)

In the 1980s, US zoologist Thomas Fritts (he of the 2002 report on George) got stuck into the wealth of data on the size and shape of tortoises from different islands. His statistical analysis revealed that Vice-Governor Lawson's boast had some truth in it: a few of the different subspecies can be reliably identified just by scrutinizing them. But most, he warns, are difficult to tell apart.

The tell-tale differences between the San Cristóbal and Santiago tortoises are subtle and would take a trained eye to spot. Lawson may have had such a trained eye; wannabe geologist Darwin certainly didn't. 'If those were the only two tortoises you had seen, they're just about the worst two you could pick to be impressed by the differences from island to island', says science historian Sulloway. So little was Darwin impressed that he not only failed to collect scientific specimens of the creatures, he apparently helped his *Beagle* shipmates tuck into the last of some 30 large tortoises during the cruise to Tahiti.

At dawn on 19 October 1835, the *Beagle* – with Darwin on board – sailed, without stopping, past Pinta and out of the archipelago. Darwin would have seen the island's impressive volcano that morning and he might have caught sight of giant tortoises in the distance. One of these could have been the female that about a century later would lay an egg containing the baby tortoise that became the world's most famous reptile.

How can DNA tell us anything about the spread of tortoises through the archipelago? The idea is that the longer two populations have been apart, the more time they have had to acquire genetic differences. Based on the assumption that such changes accumulate at a fairly constant rate, the most likely scenario is that tortoises first set foot in the Galápagos around two to three million years ago. Originally, they probably colonized either San Cristóbal or Española, the oldest islands that are closest to the mainland. At that time, the archipelago looked very different: several of the western-most islands, including Fernandina and Isabela, had not yet emerged from the ocean floor.

Eventually, tortoises were probably washed off San Cristóbal or Española and carried on the prevailing northwesterly currents. Some of these unwitting migrants presumably floated and floated, past island after island, out into the expanse of the Pacific Ocean where they eventually died and sank without trace. A few pioneers were lucky enough to be washed up on another island. The next stop was probably Santa Cruz. Once a population had established itself there, the same thing most likely happened again. Some animals from Santa Cruz were washed out to sea and survived the hop to a neighbouring island. So it was that tortoises reached Pinzón, Santiago and eventually Isabela. The tortoises then spread along the length of Isabela, ultimately forming isolated populations on each of its five volcanoes.

In sum, it seems that tortoises slowly spread throughout the archipelago, making one relatively short and plausible journey at a time. That said, some mysteries in their DNA can't be ignored.

One puzzle is Wolf, Isabela's northernmost volcano. There are two fairly discrete populations – Puerto Blanco on the northern slope and Puerto Bravo on the western slope of the volcano. Most of the tortoises here have domed shells and seem to have come from the island of Santiago just 50 km away. In 2002, Adalgisa Caccone and Jeffrey Powell, co-directors of the Yale group, led a study of DNA from 22 Wolf tortoises and found that several are clearly not direct descendents of the Santiago tortoise. One Puerto Blanco tortoise appears to have come from San Cristóbal and four Puerto Bravo ones are very similar to those from Española. Looking at the map, it is hard to imagine how tortoises could have floated directly from San Cristóbal and Española to the inaccessible reaches of Wolf volcano.

Caccone's best guess is that humans are responsible. Pirates and whalers often stopped off at the southern islands to stock up on tortoises. The frequent passage of ships sailing past Wolf volcano towards fertile whaling grounds to the north and west of the Galápagos archipelago may, she says, have made this site ideal for whaling crews to deposit excess tortoises collected elsewhere to be retrieved upon their return.

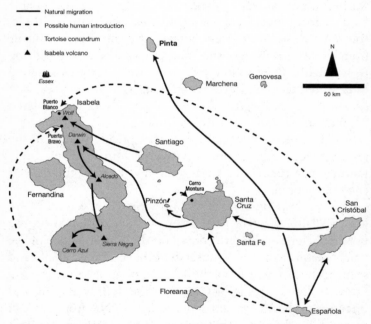

Figure 4.4 DNA evidence suggests giant tortoises arrived on either San Cristóbal or Española and spread stepwise to the other Galápagos islands

'Sail ho! Sail ho!' The cry stirred Captain David Porter from his bunk at dawn on 29 April 1813. It was to be a rewarding day for the skipper of the infamous US frigate *Essex*. During the war of 1812 – a dispute between the US and Britain over maritime law – Porter ruled the waters around the Galápagos, seizing no fewer than 15 British whaling vessels. That April morning, he came to the bridge and swept the horizon with his telescope until he had the strange sail in his sights. It was a British whaler – the *Montezuma*; Porter ordered his crew to give chase. By nine o'clock that morning, he had captured the hapless *Montezuma* with its 1400-barrel bounty of whale oil and was interrogating its captain about two boats in the distance. These, he was told, were two other British whalers, the *Georgiana* and the *Policy*. Porter gave chase once more.

Figure 4.5 US frigate *Essex*

'At eleven A.M., according to my expectation, it fell calm; we were then at the distance of eight miles from them', he wrote in his *Journal of a Cruise* of 1815. He learned from the captain of the *Montezuma* that the *Georgiana* had 35 men on board and carried 6 large guns; the *Policy* had 26 men and 10 smaller guns. In spite of the risk, the confident Porter was 'determined to have them at all hazards'; he sent out several rowing boats loaded with armed men.

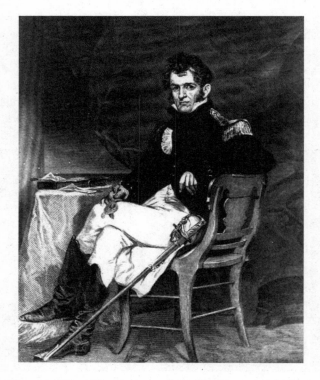

Figure 4.6 David Porter, the swashbuckling captain of the *Essex*

Those on board the *Georgiana* and *Policy* were taken by surprise. If the rowing boats reached them, they would have to fight at close quarters and casualties could be heavy. With no wind, there was no escape. Their only hope was to blast the small boats out of the water. They were woefully unprepared, particularly as they had just stocked up on tortoises, which were getting in the way. 'In clearing their decks for action, they [threw] overboard several hundred Galápagos terrapins', wrote 12-year-old *Essex* midshipman David Farragut in his diary. Farragut was in one of the rowing boats closing on the *Georgiana*:

> The appearance of these [land] turtles in the water was very singular; they floated as light as corks, stretching their long necks as high as possible, for fear of drowning. They were the first we had ever seen, and excited much curiosity as we pushed them aside.

As Farragut and his fellow men picked their way through a sea of giant tortoises, the cannon fire began. Back on the *Essex* at a safe distance, Porter looked on: 'At two o'clock, the boats were about a mile from the vessels ... when they hoisted English colours, and fired several guns. The boats now formed in one division, and pulled for the largest ship, which, as they approached, kept her guns trained on them.' None of the canons hit their mark. The boats 'rowed up beneath the muzzles of the guns and took their stations for attacking the first ship', Porter explained in a letter back to his superiors in Washington. The British struck their flag and stood down without a shot being fired. They then left a crew on board and took their stations for attacking the other vessel, the *Policy*; her flag was also struck.

'Thus were two fine British ships ... surrendered, without the slightest resistance, to seven small open boats, with fifty men, armed only with muskets, pistols, boarding-axes and cutlasses!' Porter boasted 'that Britons have either learned to respect the courage of the Americans, or they are not so courageous themselves as they would wish us believe'.

As dawn broke a couple of days later, the *Essex* with her three prizes in tow was surrounded by about 50 live giant tortoises. They were rounded up and brought on board. According to Porter's journal, 'they had been lying in the same place where they had been thrown over, incapable of any exertion in that element, except that of stretching out their long necks'. He and his crew ate these tortoises in the coming months, but presumably some of those jettisoned by the embattled British ships were never recaptured. One or two, perhaps, made it to the nearest shore. We know from Porter's letter to Washington that this would have been the northwest coast of Isabela, from which a beached tortoise could have made it to either Puerto Blanco or Puerto Bravo.

This story neatly illustrates how human activity could have contributed to the movement of tortoises from island to island.

Elsewhere in the archipelago, DNA has been able to solve a long-standing mystery. On the northwest of Santa Cruz at a place known as Cerro Montura, there is a tiny population of tortoises, which look unlike either of the island's two other populations. While most of the Santa Cruz reptiles have dome-shaped shells, the Cerro Montura tortoises are saddlebacks. How come? Analysis of the saddlebacks' genes reveals a striking similarity to tortoises from nearby Pinzón, implying that the Cerro Montura animals could be the only survivors of the population that colonized Pinzón around a million years ago. Alternatively, and probably more likely, this could be another example of human meddling. Perhaps whalers took a few tortoises from Pinzón and let them off on the northern shores of Santa Cruz.

DNA throws up an even greater puzzle though. Whilst most of the islands seemed to have been colonized in stepping-stone fashion, there is one notable exception – Pinta. As the only surviving member of the Pinta population, Lonesome George's genes suggest something quite extraordinary. Before the Yale group delved into his genome, it was widely assumed that tortoises had reached Pinta from the northern tip of Isabela. This, after all, is the island closest to Pinta that has tortoises. But Lonesome George's genes make this unlikely. His genetic sequence turns out to be almost identical to that of an Española tortoise.

Remarkably, a physical resemblance between the Pinta and Española tortoises had been realized more than 100 years earlier by William Cookson, a British captain who collected three tortoises from Pinta in 1875. In a prescient letter back to tortoise buff Albert Günther at the British Museum, Cookson wrote: 'As Hood [Española] and Abingdon [Pinta] Islands are the most distant from one another of any in the group, it would have been very interesting to ascertain if their tortoises really resembled one another, whilst they differed from those of the other islands.' With his time in the archipelago limited, Cookson decided to look for tortoises on the two islands where his chances of success were highest – Isabela and Pinta. He never saw an Española tortoise, so never made his 'interesting' comparison.

The Yale geneticists did. Adalgisa Caccone, team co-director, didn't believe what they found. Worried that the blood sample from Lonesome George had somehow been contaminated, Caccone put the analysis on hold until she could get more blood. The following year, she extracted DNA from a fresh sample and reran the analysis. The result was exactly the same. Genetically speaking, Lonesome George is virtually indistinguishable from an Española tortoise. Rather than making the relatively plausible 100-km journey from the northern tip of Isabela to Pinta, Lonesome George's DNA suggests that his buoyant ancestors floated more than 300 km from the very south to the very north of the archipelago.

Caccone thinks ocean currents could be the secret. 'There is a strong current running northwest from the northern coast of San Cristóbal leading directly to the area around Pinta', she and her colleagues wrote in *Proceedings of the National Academy of Sciences* in 1999.

There's some evidence that another group of animals – the lava lizards – made a similar odyssey from the south to the north. Taxonomists – our life librarians – recognize at least seven species of the small and fairly ordinary Galápagos lava lizard. Back in 1983, protein analysis suggested that the lizards on Pinta were most closely related to those on southerly Floreana. More recent work using current molecular techniques presents a more complex picture. But it doesn't rule out the possibility of a leap, just like the one that Caccone proposes for the origin of the Pinta tortoise.

This kind of long-distance migration may be an exception to most of the animal movements within the archipelago, but it's a lot less extraordinary than the colonization of the Galápagos in the first place.

—

There is another, more alarming, explanation for the incredibly close genetic similarity between Lonesome George and Española tortoises: Lonesome George *is* an Española tortoise. Perhaps the Pinta population really is extinct and the animal found in 1972 is a recent visitor that somehow found his way from island to island.

This argument is difficult to test. But it's not impossible. A handful of tortoises that we know were collected on Pinta in the 19th and early 20th centuries wound up in museums. If Lonesome George is a bona fide Pinta animal, then his DNA sequence should match that of these other specimens.

Ed Louis is a geneticist at Henry Doorly Zoo in Omaha. When he tried to extract DNA from three specimens collected on Pinta in 1906 that were killed, skinned and are now housed in the California Academy of Sciences in San Francisco, he reached an uncomfortable conclusion. Despite repeated efforts, only two of the three tortoises yielded any DNA and the sequence from these two samples did not match that of Lonesome George. This suggests that our celebrity tortoise could be an introduction, Louis says.

Worrying stuff. Lonesome George's roots are very important. He is only of special scientific interest if he really is a Pinta tortoise. If someone just dumped him there a year or two before he was found, he doesn't warrant his own enclosure or a single column-inch of the media attention he has attracted over the years. If Lonesome George is an impostor from another island posing as a Pinta tortoise, he has fooled millions.

This same concern occurred to Caccone and her colleagues. They too got hold of skin samples from the three tortoises at the California Academy and managed to extract enough fragments of DNA to compare against Lonesome George's sequence. Matching up DNA fragments is rather like trying to do an incomplete jigsaw. If you have very few pieces, it is difficult to see the overall picture – to realize (say) that the dome and minarets are from the Taj Mahal. If you have almost all the pieces, you can be pretty certain.

The Yale group recovered about three-quarters of the genetic sequence from the museum specimens, enough DNA to be confident that Lonesome George is from the same population as the other animals collected from Pinta. It's not obvious why Louis failed where Caccone succeeded, but getting genetic material from a specimen that's been dead for nearly a century is far from easy.

On balance it looks like Lonesome George fully deserves his hard-earned celebrity status.

Chapter 5
MAN TRAP

Lonesome George is a poster boy for conservation in the Galápagos and beyond. He is what conservation experts call a 'flagship'.

Flagship species are charismatic animals or occasionally plants that stimulate conservation awareness and action. A successful flagship need only operate in the world of marketing and public relations. It need not even be endangered if it achieves something for conservation.

Anyone can go to the Galápagos, if they've got enough money that is. It's the world's premier ecotourism destination, now attracting over 100,000 visitors each year. Most of the tourists who arrive in Baltra have heard about Lonesome George and will take a morning or an afternoon out of their trip round the islands to visit him at the Charles Darwin Research Station (CDRS) in Puerto Ayora on Santa Cruz.

Every day, coloured signposts point hundreds of tourists down custom-made walkways to a few small enclosures – home to some of the archipelago's most endangered animals. The brightly clad visitors crouch respectfully beside captive tortoises; the only sounds are a rhythmic munch from the animals, the pregnant click of camera shutters and soft words from tanned guides in khaki shorts.

Each guide leads their group onwards to the succession of pens that house tortoises from hatching until they are old enough to be returned safely to their own islands. And at the end of the trail through the research station, the tourists get a few minutes to commune with Lonesome George.

They've all heard about him; this is the highlight and occasionally the reason for their visit to the station. The final information panel on the platform connects him explicitly with the conservation message: 'Whatever happens to this single animal, let him always remind us that the fate of all living things on Earth is in human hands.' He's a flagship alright.

What makes a good flagship? Top of the hit parade are mammals; birds occasionally get a look in, reptiles rarely, fish hardly and invertebrates can forget it. The spineless just don't have one of the most important qualities of a good flagship – public appeal.

There is something going on in this hierarchy. Not only do mammals have fur, so can easily be rendered cute and cuddly, they also have facial features that are not dissimilar to ours. Humans respond positively to fluffy, anthropomorphized animal caricatures. This explains why bears, monkeys and lions – even if not particularly endangered – are among the most enduring flagship species. It is also the reason why many fictional animals, such as Simba, Disney's Lion King and Pixar's clownfish Nemo, may qualify for flagship status; the human qualities conferred on these cartoon creations makes them vastly appealing to an anthropocentric public.

Nevertheless, a fair few flagships are harder to account for. The spindly legged, slender-billed avocet isn't particularly cute or easy to relate to but is the emblem for the UK-based Royal Society for the Protection of Birds (RSPB), the largest wildlife conservation charity in Europe. It suggests that other qualities can lead to flagship status.

Up until the 19th century, breeding avocet were a common sight in the marshes and fens of eastern England. By 1842, they stopped breeding in the region owing to extensive draining of land to reclaim it for agriculture. It was only during the Second World War that military defences recreated the flooded conditions that this species favours. In

1947, avocets returned to breed in England, the RSPB offered them protection and they have been on the increase ever since. Today, between 600 and 800 pairs breed along the east coast each year. The avocet isn't cute, but its conservation message is strong.

If these two qualities – public appeal and a conservation message – come together in one species, you've more than likely found a powerful flagship. The giant panda is the perfect example. Most captive individuals have names and are treated as honorary humans by zoo staff and visitors alike. The panda is also highly endangered. This winning combination has made giant pandas the star attraction at zoos the world over for more than 50 years.

Indeed, a few individual pandas have become flagships in their own right. When Richard Nixon made the first US presidential visit to China in 1972, the Chinese government gave the American people two pandas – Hsing-Hsing and Ling-Ling. They became a symbol of the new friendship between the countries and were a huge hit at the National Zoo in Washington DC. On their first day, they drew crowds of more than 20,000; in their lifetime, it's estimated that more than 75 million people flocked to see them. 'They represented the world of nature', wrote zoo staff after the death of Hsing-Hsing in 1999. 'Their status as one of the world's most endangered animals, coupled with their popular appeal, helped introduce millions of people to conservation and brought the message of a shrinking wilderness and threatened species to an urban population.'

Perhaps the greatest flagship individual was Chi-Chi, a panda that arrived at London Zoo in 1958. She had been destined for an American zoo but this was before Nixon's pandas thawed the diplomatic tensions between China and the US, and as a symbol of communism she was refused entry. London Zoo decided to give Chi-Chi asylum and she was an instant hit with the public. So much so that she was the inspiration for the World Wildlife Fund's striking panda logo. Chi-Chi is now immortalized as the symbol of global conservation.

Only a few individual animals have become brands in this way. Elsa, the lion cub reared in captivity and subsequently set free by Joy Adamson, had the same combination of public appeal and a message. Once she had inspired the book, the film and the marvellous *Born Free* soundtrack, she went on to become the emblem of the Born Free Foundation.

While we're at it, the story of Brighty the Grand Canyon donkey is worth telling just to prove that flagship species don't need to be remotely endangered. During the mining boom of

Figure 5.1 A statue of Brighty the Grand Canyon donkey receiving a pat from National Park Superintendent Howard Strickland

the late 19th century, the canyon's miners brought donkeys with them to carry out their precious excavations. When they packed up their pick axes and tourism took over, they set the donkeys loose. By 1919, when the canyon achieved national park status, these introduced animals had become a real tourist attraction. One of them – a friendly, pancake-eating, crowd-pleasing beast called Brighty – soon achieved celebrity status, inspiring people to visit and value the canyon. He became the premier icon of the American wilderness, according to John Wills, who specializes in US environmental history at the University of Kent in the UK. 'Brighty provided a way to negotiate the landscape', he says.

Sadly, Brighty disappeared during the winter of 1922, shot and eaten (as legend has it) by two starving travellers who failed to recognize the iconic burro. Nevertheless, he became the subject of a series of best-selling books published in the 1950s and a shiny-nosed bronze statue of him sits in the entrance hall of the Grand Canyon Lodge on the northern rim of the canyon.

Like Chi-Chi, Elsa and Brighty, Lonesome George is a celebrity with the double whammy of charisma and a conservation message. His one-of-a-kind condition and his reclusiveness are things people can relate to – his story triggers a sympathetic response. And the fate of his subspecies echoes the fragility of endangered living things everywhere. What makes his flagship status particularly impressive is that he is not a mammal, but a reptile. There are very few reptilian flagships – turtles focus the mind on marine protection and the Komodo dragon draws ecotourists to its eponymous Philippine island – but these are species rather than individuals. What's more, Lonesome George has become the face of not one organization but many – the Charles Darwin Foundation, the GNPS and a string of overseas charities such as the UK-based Galapagos Conservation Trust.

Figure 5.2 Logo of the UK-based Galapagos Conservation Trust

The first tourist vessel – a 66-passenger ship from Chile – entered Galápagos waters in 1967. The subsequent growth of this new industry is probably the most significant of several factors that have changed the islands forever, making this province one of the fastest growing and most affluent in an otherwise impoverished Ecuador. It's an incredible transition, because for most of their history, these isolated and inhospitable islands were not a particularly desirable destination.

When Herman Melville passed through in 1841 gathering material that ended up in his blockbuster novel *Moby Dick*, he wrote a series of 10 sketches for *Putnam's Monthly Magazine* under the collective title 'The *Encantadas* or, Enchanted Isles'. He didn't mean they were enchanting. Far from it. 'A group rather of extinct volcanoes than of isles, looking much as the world at large might after a penal conflagration', he wrote. For Melville, as for others before him, these were haunting islands.

The first person brave (or crazy) enough to think about living there was an Irishman, Patrick Watkins, marooned on Floreana in 1807. He survived alone for several years, growing vegetables and selling them to passing ships or exchanging them for rum. The swashbuckling Captain Porter of the US frigate *Essex* recounted in his 1815 *Journal of a Cruise* how Watkins was faring with the rigours of Galápagos life:

> The appearance of this man, from the accounts I have received of him, was the most dreadful that can be imagined; ragged clothes, scarce sufficient to cover his nakedness, and covered with vermin; his red hair and beard matted, his skin much burnt, from constant exposure to the sun, and so wild and savage in his manner and appearance, that he struck every one with horror.

Whaling vessels like the *Georgiana* and *Policy* captured by Porter were a pretty common sight in Galápagos waters. Only the eccentric Watkins felt the urge to spend any serious time there.

In 1832, just before Darwin passed through, Ecuador annexed the Galápagos. This triggered the first serious attempts to colonize the archipelago. The first governor general of the islands, the wealthy General José Villamil, had ambitious plans. He wanted to harvest and export the abundant lichen *Roccella tinctoria*, a good source of a purple dye called orchil. He failed. In 1870, another entrepreneur, Ecuadorian José de Valdizan, had another go; this too came to nothing. In 1878, his workers rebelled and murdered him.

At the beginning of the 20th century, there were still only around 600 inhabitants in just a few settlements. The archipelago remained similarly uncivilized until the middle of the 20th century. During the war, a US military base appeared on Baltra, giving the islands their first runway. Then, in 1959, Ecuador declared about 97% of the land a national park. The first tourists arrived just eight years later, marking the beginning of a new phase in the history of the islands.

Tourists mean money. The Galápagos had found a reliable and relatively lucrative industry that could sustain a larger population. There was no looking back. The number of ecotourists visiting the islands and the size of the resident population required to service them have continued to rise hand in hand.

This in spite of concern over the number of tourists that the islands can sustainably hold. The first management plan for the national park in 1974 recommended that tourist numbers should not rise above 12,000. By 1979, this notional limit had been exceeded. At the end of the 20th century, some 80,000 tourists were entering this fragile, unique ecosystem every year. Although these ecotourists are strictly instructed to 'leave only footprints', the reality is, of course, very different.

One of the single biggest long-term consequences of humans coming and going is the introduction of non-native or 'alien' species. Taking a plant or animal out of its natural setting to another place frees it from its natural web of pests, predators and competitors for food and space. Thus unencumbered it can be unstoppable. Hence the inadvertent and occasionally deliberate introduction of alien species has been a disaster for local wildlife and even entire landscapes in the Galápagos, as elsewhere.

In the late 19th century, an organisation known as the American Acclimatization Society fell upon the twee notion of bringing all the birds mentioned in Shakespeare's works to North America. 'Nay, I'll have a starling shall be taught to speak Nothing but "Mortimer", and give it him To keep his anger still in motion', said Hotspur in Shakespeare's *Henry IV*. So in 1890, chief acclimatist Eugene Schieffelin set free 60 European starlings in New York's Central Park. These foreigners thrived, but at the expense of native birds like cardinals, chickadees and finches – today there are reckoned to be over 200 million European starlings across North America. Thankfully for North American birds, Schieffelin's efforts to import other Shakespearean birds like bullfinches, nightingales and skylarks were not so successful.

Figure 5.3 European starlings, introduced to the US by literary aspirant Eugene Schieffelin

Australia is well known for disastrous introductions. First came the rabbit, a perfect example of how the sudden appearance of an apparently benign creature can wreak havoc on a finely balanced ecosystem. Rabbits arrived in Australia in the middle of the 19th century, introduced by wealthy Europeans to fulfil their desire to hunt. The rabbits took to their new home and at first everyone was delighted. But as Australia witnessed the fastest spread of any colonizing mammal anywhere in the world, the landscape soon became covered by a 'grey blanket'. The bunnies razed vegetation and caused untold damage to precious crops; they pilfered the food and annexed the burrows of native marsupials like the greater bilby and the burrowing bettong. Although the rabbit-specific *Myxoma* virus released in 1950 did a pretty good job at bringing down numbers, their resistance to this control measure has slowly increased. Today, the annual cost of production losses and control measures is somewhere around a million US dollars.

Then came the cane toad *Bufo marinus*, introduced from the Americas. In 1935, 101 toads were released in Queensland to control the greyback beetle that was devastating sugar cane crops. This wasn't quite thought through either. Anything that eats a cane toad gets a nasty surprise: poison is released from glands behind its head. Red-bellied black snakes were hit hard, and with a decrease in the abundance of red-bellied black snakes, there was an explosion of rodents and grasshoppers. Birds like the pheasant coucal, white-faced heron and eastern swamp hen have all been disturbed by the toad; and native amphibians such as the white tree frog are snaffled up by the non-native. Australian scientists continue to work on ways to prevent the southern advance of this species.

The latest alien tragedy for Australia comes in the unlikely form of an ant. No ordinary ant, the South American imported red fire ant *Solenopsis invicta* delivers a venomous sting and is threatening to do severe damage to wildlife on the eastern coast of Australia. They cause trouble for native ants; they eat land and freshwater snails and sting all sorts of amphibians; they devastate clutches of turtles, lizards, snakes and crocodiles and happily feed on newly hatched birds.

Animal introductions to the Galápagos have been equally if not more troublesome. Large mammals like pigs and goats have had the most obvious impact, more of which anon. In addition, several less conspicuous mammals, a handful of reptiles and birds, scores of invertebrates and hundreds of plants have reached the archipelago with a helping hand from humans. Some of these have had a dramatic effect on native species. What follows is just a taster.

Black rats probably made it to the archipelago on ships as

long ago as the late 17th century. On every island infested by black rats, the native rice rats have gone extinct. Every island that is except Santiago, where the small rice rat population was rediscovered in 1997. Donna Harris, a doctorate student at Oxford University, reckons that the Santiago native has hung on because it copes better than the invader during this island's frequent droughts.

The parasitic fly *Philornis downsi* probably arrived in the islands some time in the 20th century. Until 2003, it was documented in only two places – Santa Cruz and Isabela. A recent survey found that it is much more widespread and is probably now common on most islands. This is bad news for the native birds. The fly lays its eggs in the nests of at least 15 species including Darwin's finches, mockingbirds and flycatchers; when the larvae hatch out, they burrow into the flesh of the chicks and feed off their blood. One in five finch clutches is destroyed by these flies. An insecticide could be used to fumigate infected nests, but this is labour-intensive. Field researchers would have to locate nests, remove the chicks and apply the insecticide. Nevertheless, this kind of intervention might be justified for an endangered species like the mangrove finch.

The red quinine tree *Cinchona pubescens* is one of the most serious alien plant invaders in the Galápagos. First planted in the agricultural zone of Santa Cruz in 1946, its vigorous growth and rapid reproduction enabled it to spread over a vast area of the highlands. Now quinine is significantly reducing the native vegetation. Fortunately, a chemical cocktail pasted into machete cuts on the trunks of these trees works wonders. This 'hack-and-squirt' approach to quinine forests on Santa Cruz is now standard procedure for the national park wardens.

In rare circumstances, the best way to tackle an alien is to introduce another – often its natural enemy – to control it. In 1982, scientists found a non-native insect pest, the cottony-cushion scale *Icerya purchasi*, sucking the sappy goodness out of all manner of plants on San Cristóbal. It has now spread to at least 14 other islands. Elsewhere in the world, infestations of this pest are managed by bringing in its predator, the Australian ladybug beetle *Rodolia cardinalis*. Why not in the Galápagos too?

As the story of the cane toad demonstrates, introducing one alien to kill another can create an even bigger problem. Galápagos scientists and managers had to be certain that the ladybugs weren't going to start eating up a whole bunch of native species. They conducted careful feeding experiments, which revealed the ladybug to be rather fussy, concentrating its efforts almost exclusively on the cottony-cushion scale. It did not pose a threat to several native insects that could have been potential targets. In January 2002, CDRS scientists and national park wardens released ladybugs onto four islands and later onto five more. The bugs have since dispersed, reaching at least another three islands. Initial signs are that this radical measure is having the desired effect. 'The mangrove stands of Puerto Ayora, previously blackened and dying from the effects of the cottony cushion scale are now green and thriving', reported a 2005 issue of *Galápagos News*.

Another direct consequence of booming tourism could be that the natural behaviour of the animals starts to change. If this happens, then it does so over a long period and is going to be rather difficult to detect. On Floreana in 1835, Darwin watched a boy with a cane killing doves and finches as they came to drink at a well. 'He had already procured a little heap of them for his dinner, and he said that he had constantly been in the habit of waiting by this well for the same purpose', wrote Darwin in his *Journal of Researches*. It would take many generations facing this kind of persecution for such creatures to evolve an instinctive suspicion of humans, he suggested.

Ecotourists are generally pretty well behaved. With only occasional cajoling from the natural history guides who must accompany them, they stick to well-defined paths and dive sites and certainly don't go around killing animals. Consequently, the wildlife in the Galápagos is still surprisingly tame. Indeed, much of the island fauna is so comfortable with the presence of humans that the absorbed tourist gawping into the middle distance runs the very real risk of stepping on an iguana or a blue-footed booby. The iguana will simply snort and the booby will let out an alarming honk.

Figure 5.4 Tourists in Puerto Ayora

There could, however, be rather inconspicuous changes taking place. Martin Wikelski, an evolutionary biologist at Princeton University, compared levels of the stress hormone corticosterone in marine iguanas at a tourist site with levels in iguanas living away from the tourist trail. At capture, there was no difference in corticosterone concentrations. Interestingly, there was a difference in the stress response following capture. Iguanas exposed to tourists had a much smaller increase in corticosterone levels than those unaccustomed to people. If anything, it looks like the endless passage of tourists may have made the iguanas too chilled out. In a real emergency, perhaps they might not respond as urgently as evolution intended, Wikelski and a colleague suggested in the journal *Biological Conservation* in 2002.

Even if there are such subtle consequences on animal physiology and behaviour, tourists still only visit a small area of the national park. The number of animals that rarely encounter humans vastly outweighs those exposed to a steady stream of shorts and sandals.

In just 30 years, the human population in the Galápagos has increased tenfold. In the 1990s alone, the population on central Santa Cruz doubled to over 10,000. In 2005, it was estimated the total population topped 27,000. The infra-structure needed to support this population expansion inevitably increases development, putting pressure on natural resources like timber, minerals, fossil fuels and water. Managing these needs in a sustainable fashion is hard.

In 2002, building started on the last urban area on Santa Cruz that is assigned for housing. What will happen when more settlements are needed? There is room for expansion into the underdeveloped farmland, but could there also be pressure on the Galápagos authorities to eat into the national park? Already, more people are deciding to move to Isabela, a relatively uncluttered island with plenty of land outside the park.

Construction has increased the demand for materials. Loggers weighed into large specimens of native trees like the

Matazarno *Piscidia carthagenensis*, prized for its strength and resistance to rot and insects. The consequences for the delicately balanced ecosystem are unclear. But with returns diminishing, the loggers turned to other native species like the Guayabillo *Psidium galapageium*.

The number of imported cars and trucks is rising. In the 1970s, there were just three vehicles on Santa Cruz; now the town buzzes with more than 300 taxis. In time, this may mean more roads, fragmenting habitats, spreading alien species and increasing road kills. Several hundred birds die on Santa Cruz's roads every week. The cones of fine lava that are quarried for building projects contain rare snails (and other inconspicuous creatures), sometimes the world's only population dwelling on one cone. Pulling down these mounds to pave the way for the expanding population could wipe out entire species.

All this development means the demand for energy is spiralling out of control. At the moment, this is met by shipping in fossil fuel from the mainland. In 2002, around 6 million gallons of diesel were brought to the archipelago. Transportation by sea, in boats not built for the job, is less than ideal.

The main towns in the Galápagos are not blessed by deep-water docks. Tankers are supposed to siphon off their cargo at a central terminal on Baltra or remain at sea and unload into awaiting barges. Most of these go-between vessels that distribute the fuel to the towns around the archipelago are not designed to navigate in rough seas or carry such a sensitive cargo. Small spills occur all the time and big spills are inevitable.

Just before 10 p.m. on 16 January 2001, the *Jessica*, an oil tanker, was approaching Puerto Baquerizo Moreno, the administrative capital of the islands on San Cristóbal. Mistaking a signal buoy for a lighthouse, the captain steered the tanker onto a rocky reef in the appropriately named

Wreck Bay. Deep in the hold sat 160,000 gallons of diesel fuel bound for the terminal on Baltra and a further 80,000 gallons of the much thicker and stickier bunker fuel to be delivered to the tourist cruise ship the *Galápagos Explorer II*.

A couple of days later, fuel began to spill from the *Jessica*'s hull. The delay between the grounding and the spill was fortunate. It allowed GNPS wardens and CDRS scientists to conduct a quick survey of the flora and fauna at key locations that were expected to suffer should the tanker release its cargo. The Ecuadorian navy and US coastguard recovered some of the fuel, but most leached into the sea where currents carried it towards Floreana, Santa Fe and Santa Cruz. The devastation was not as serious as it could have been. The prevailing northwesterly current and warm weather conditions dispersed much of the slick and minimized its impact.

Then there's freshwater, which has always been difficult to come by. Three days before Bishop Fray Tomás de Berlanga reached the archipelago in 1535, his ship's supply of water ran out. 'All of us, as well as the horses, suffered great hardship', he wrote in a letter to King Charles V of Spain. Once onshore, some men began to dig a well and others made off into the interior looking for water. 'From the well there came out water saltier than that of the sea; on land they were not even able to find even a drop of water for two days', Berlanga told the king. Things soon got desperate: 'with the thirst the people felt, they resorted to a leaf of some thistles like prickly pears, and because they were somewhat juicy, although not very tasty, we began to eat of them', he wrote. Berlanga and his men eventually – with what they took to be God's help – found water in a ravine. This did not stop two men and 10 horses dying from dehydration.

From a few spartan clues in Berlanga's letter, historian John Woram has worked out that this island was most likely Floreana, now known for its reliable source of good-quality water. Had the bishop landed on another island where water is less easily accessible, he and his men might not have survived to tell the tale.

If Berlanga struggled to find water on Floreana, just imagine the challenge of meeting the needs of around half the archipelago's population that lives in Puerto Ayora. Seven electric pumps draw up water from deep fissures where rain and sea water mix. In an El Niño year, there is so much rain that the water is close to fresh. But on the whole, it's rather salty and is really only suitable for domestic chores. Drinking water mainly comes from rainwater collection tanks that sit beside some buildings or from bottles imported to the islands.

Agriculture is also a problem. Development in the archipelago is restricted to just 3% of its land area. The need to grow crops and graze cattle still places a huge pressure on that 3%. The best place for agriculture and livestock farming is in the lush, humid highlands. Here rainfall is highest and water is relatively plentiful, and on all the islands where humans have settled, agriculture is permitted in much or sometimes all of this zone. Unfortunately, the conditions that make this area best suited for agriculture also make it the most species-rich. So clearing highlands to make way for farming had devastating effects. San Cristóbal, for example, lost all its humid zone (and presumably countless unique species) to agriculture.

This kind of damage was mostly done before the 1980s. Since then, the area of land under agricultural use has declined slightly. This isn't necessarily a good thing. Much former farmland now lies abandoned, as people flood to town to benefit from more lucrative jobs in tourism or fishing. Such unused open spaces are a gift to non-native species, encouraging the spread of introduced plants and animals. Less farming also means a greater dependence on importing produce from the mainland, and with more imports come more aliens.

The present rate of population growth – doubling every decade or so – is unsustainable, says Graham Watkins, current director of the CDRS. If this pattern continues for much longer, it will become impossible to manage its impact.

The Special Law for the Galápagos, passed by the Ecuadorian Congress in 1998, recognized the need to slow immigration. There are now only three main ways to qualify for permanent residency: if you are born there, if you lived there for five or more years before the law or if you are the spouse or child of someone who already has residency. Ironically, these restrictions may have stimulated a rush to the archipelago as Ecuadorians seized what they saw to be their last chance to move to the Galápagos. In time, hopefully the population expansion should begin to tail off, reducing the pressure on the islands' finite resources.

The Special Law also precipitated action against introduced species. Obviously the best way to tackle these outsiders is to stop them arriving in the first place. There are now regular inspections of luggage in the arrival hall at Baltra airport and boat cargo on the dockside in Guayaquil where most of the supply vessels originate. Still, airport inspections during 2002 revealed that although tourists on aeroplanes outnumber locals by 2 to 1, it is the locals who bring in around three-quarters of introduced species. This highlights that further education is needed to get the message across.

Education and better legal protection are all part of a new vision for the Galápagos, one in which its inhabitants appreciate the constraints of island living. Much that goes on on the mainland simply won't work in such a remote location, says CDRS director Watkins. The limited availability of natural resources, difficulty establishing economies of scale and the huge distance to the mainland put the national and international markets well beyond the reach of many Galápagos businesses.

Those that produce low-volume, high-value goods can just about survive. For example, coffee is now being grown from trees planted on San Cristóbal by French explorers in 1869 and exported as a specialty product. A really sustainable Galápagos business is one that can tap into this kind of low-impact niche market, says Watkins. The alternative is to resist the temptations of globalization and focus on what is going to sell in local or tourist markets, he says.

Some small businesses are now taking this approach, making value-added products from local produce, like jams, fish pâtés and wooden handicrafts carved from introduced tree species. Farmers too are being encouraged back to abandoned agricultural land and to concentrate on crops for local consumption.

There is talk of doing away with fossil fuel imports by turning the archipelago towards renewable and low-polluting energy sources. The small town of Puerto Velasco Ibarra on Floreana is experimenting with solar power. The GNPS has now installed donated solar panels to power, among other things, the incubation of giant tortoise eggs at the CDRS and at Puerto Villamil's tortoise-rearing centre on Isabela.

Unsurprisingly, of the many threats to Lonesome George's world, these subtle issues of land use and ecosystem integrity get the least press. What really hits the headlines is the plundering of the Galápagos waters.

Chapter 6
LOCK UP YOUR TORTOISE

'¡*Muerte al Solitario Jorge!*' – 'Death to Lonesome George!'
chanted the *pepineros*. In January 1995, when the sea
cucumber fishermen took over the offices of the Charles
Darwin Research Station (CDRS), they threatened to
butcher its animals. The emblematic Lonesome George was
the target of their most violent threats. He had unwittingly
become a pawn in an ongoing tussle between conservationists
and fishermen.

In Asia, sea cucumbers are prized as an aphrodisiac. Until
recently, populations along the west coast of South America
fed this market. During the 1990s, dwindling supply failed to
meet increasing demand and the going rate for a *pepino de mar*
began to escalate. At the end of the 1980s, a kilogram of dried
sea cucumbers would have set you back around $10. Within a
few years, this had more than doubled. Fishermen only saw a
fraction of this money – the rest being creamed off by
middlemen and retailers – but for those on the breadline, this
slimy creature meant serious cash.

The problem was where to find them. The Galápagos was
the obvious place. In the early 1990s, fishermen began to
stream across to the archipelago from the continent to ransack
the rich waters around the islands. Their sights were set on
one species, the increasingly lucrative *Isostichopus fuscus*.

Sea cucumbers are not really cucumbers, they just look a lot
like them. They are in fact most closely related to starfish and
sea urchins, belonging to a group of animals called echino-
derms. Well over a thousand species have been described,

although this is probably only a fraction of all the sea cucumbers living in the depths of the ocean. There are several species in the Galápagos, but only *I. fuscus* is big and chewy enough to be worth exploiting commercially.

Figure 6.1 The much sought-after Galápagos sea cucumber

More than 10 years before Lonesome George received his first credible death threat, the Ecuadorian government decided to clamp down on foreign fishermen entering the Galápagos to exploit its rich marine resources. The Galápagos Marine Resources Reserve was born in 1986. It took until August 1992 for the government to settle the details of how the new reserve would be managed. In theory, it put the waters surrounding each island out of bounds to commercial fishing. In practice it did no such thing.

With just a few patrol boats and no aeroplanes or helicopters at its disposal, the Galápagos National Park Service (GNPS) was woefully ill-equipped to police the 70,000 km^2

reserve. International fishing vessels, mostly feeding Asian markets, continued to swarm around the islands, frequently entering the notional reserve.

In the Galápagos, *I. fuscus* lives in fairly shallow waters, inches along the bottom like a slug and doesn't bite. Consequently, the fishing technique is pretty straightforward: at its simplest, animals are harvested from the bottom by someone wading in the shallows; at its most complicated, sea cucumber fishing involves scuba diving. The most common approach is for divers to attach a snorkel to a hosepipe and scour intermediate depths without having to surface too regularly. Some species do eject sticky tubules from their anus to immobilize prey and put off predators. But the fishermen were – are – undaunted.

Godfrey Merlen is a bearded, bespectacled, sandal-wearing naturalist and boat captain who has been in the Galápagos since the early 1970s. If anyone can communicate the complexity of the sea cucumber crisis, it's Merlen. Writing in *Noticias de Galápagos* in 1993, he said:

> If I had been born in the Guasmo of Guayaquil, into the abject poverty that occurs there, into a world of harsh survival, into a world without trinkets and fancy toys such as television, Betamax, and gaudy clothes, I would jump with glee to be offered ten thousand Sucres a day to pick animals from the sea floor, to be able to join the wealthy elite gaining the power to buy my own baubles and vodka and Nike shoes.

Sea cucumbers were, literally, rich pickings. Michael D'Orso summed it up nicely in his 2002 book *Plundering Paradise*: 'As the islanders saw that a three-man crew could make as much as several hundred dollars each in a single day – this in a nation where the average per capita annual income was less than $1,600 – the business exploded.'

Here, at last, was a chance to make some money. The ban

on sea cucumber fishing imposed by the new marine reserve threatened to close the cash register before their eyes. 'It is a classic case of a group of relatively poor people finally finding a resource to better their lives and a group of relatively wealthy educated people trying to deny them the opportunity in the name of protection of the environment', wrote a couple of visiting geneticists a few years later.

The fishermen challenged the ban. What was the evidence, they demanded, that a bit of fishing would have any serious effect upon the healthy sea cucumber population? Well, the catalogue of sea cucumber fisheries around the world that have collapsed because of overfishing makes dismal reading. But the *pepineros* had a point (albeit a slender one). At that time, very little *was* known about Galápagos sea cucumbers, with only very crude estimates of numbers. Without baseline figures it was hard to demonstrate the impact of fishing in these waters.

The *pepineros* became increasingly vocal. 'It started with them just sitting on the front steps of the administration building, with placards and chanting as we were going in and out', remembers Linda Cayot. In those early days, most of the protesters were locals, who otherwise lived happily alongside outsiders like US-born Cayot. One moment, they would be chanting '*fuera gringos*' – 'go home foreigners'; then they would cheerily pass the time of day. It was a bizarre situation, she says. As more fishermen arrived from mainland Ecuador, things took a sinister turn. Unrest broke out, particularly on Isabela, the stronghold of the *pepineros*.

Why all the fuss over some slimy, sea-bottom feeders? Many reasons. From an early age, we're taught to respect earthworms. They might be unattractive but they are inherently important, leaving behind a finely digested, fertile soil that benefits other species, including (ultimately) ourselves. Sea cucumbers, nicknamed 'earthworms of the sea', are a similar crucial link in the marine food chain.

Removing them could have all sorts of dire consequences for the delicate marine ecosystem.

There might be short-term benefits for a few people, but the long-term effects will be felt for generations, a point eloquently made by Galápagos-born Carlos Valle, now professor of Evolution and Environmental Sciences at the University of San Francisco in Quito. 'Hundreds, and perhaps thousands, of Ecuadorians could get a glimpse of an ephemeral paradise brought by a sudden short-lived economic opportunity and illusory prosperity, but at the price of tomorrow's misery', he wrote in a 1994 edition of *Noticias de Galápagos*. 'The entrepreneur who promoted it will be gone, his pockets full, and here we will still be, even poorer than before.'

There was also a more pressing reason to worry about this burgeoning business. Freshly picked sea cucumbers need to be processed. They are cooked onshore in large vats of boiling water, then sun-dried, either by draping them on rocks or arranging them on specially fashioned wire racks. It was Fernandina and Isabela – the islands closest to the best sea cucumber fishing – that bore the brunt of these makeshift processing plants. The *pepineros* hacked down mangroves to fuel their fires and left rubbish and human waste in their wake. An additional concern was that such illegal camps might be introducing all sorts of alien species to these relatively pristine islands.

Tensions mounted in 1994 as park wardens came across signs of systematic slaughter of Isabela's tortoises. By the end of the year, they had found the remains of over 80 animals – male, female, young, old – more than had been butchered in the whole of the previous decade. Some had just been hacked to death and left to rot.

It might have been these poachers who forgot to extinguish cooking fires on 12 April 1994. Strong winds fanned flames across the southwestern slopes of Sierra Negra on Isabela. The world looked on as firefighters drove tractors to the base of the volcano to construct breaks and helicopters doused the flames with water. Although the blaze did not threaten the volcano's tortoises, the ongoing poaching was a big concern. Park wardens took the opportunity to give more tortoises refuge at the breeding centre in Puerto Villamil. Some were carried down the mountainside on donkeys; others were airlifted by helicopter.

Although there was probably no malice in the fire, for many these events took on a political significance. 'The destruction caused by the fire and the slaughter of tortoises were evidently used to threaten the authorities', wrote marine iguana biologist Martin Wikelski in *International Forest Fire News* in 1995.

In June 1994, Ecuador's Fisheries Development Council conceded and authorized a three-month long 'experimental' fishery. At least conservation biologists would be able to see for themselves the damage fishing was causing; they would at last have some serious data to evaluate.

The experimental fishery was carefully organized. All boats had to be registered with the GNPS and all fishermen on the boats had to have a licence. Vessels from each of the three main fishing ports – Puerto Ayora, Puerto Villamil and Puerto Baquerizo Moreno – were painted in different colours to help park officials. Perhaps, if all interested parties worked together, an acceptable compromise could be reached? Some 420 fishermen were given permits to take sea cucumbers on the condition that the total catch would not top 550,000 animals.

Things didn't go quite to plan. The sea cucumber quota was exceeded within weeks and the fishing continued. After one month, nearly 1000 fishermen were collecting sea cucumbers. About half had no permit. After two months, the catch reached a staggering 10 million animals. An inspection in early December 1994 revealed that almost every boat was breaking some park rule or other. In addition to those vessels without a permit, some were harvesting other protected species like starfish, seahorses and turtles, others had substandard health and living conditions, and much of the fishing gear was illegal. The fisheries authority stopped the season one month short. The *pepineros* – now dominated by newcomers from mainland Ecuador – were not happy.

On the morning of 3 January 1995, between 20 and 30 *pepineros* set up a blockade on the road leading to the offices of the GNPS and CDRS, demanding that the fishery be reopened. For three days, they stopped anyone entering or leaving and prevented tourists touring the research station grounds.

One of these tourists was young Ecuadorian biologist Verónica Toral-Granda. Visiting the islands with a friend, she found the road to the research station barred by masked men wielding clubs and machetes. It left an impression. Years later, Toral-Granda became intimately involved in the crisis, carrying out the first detailed research on the Galápagos sea cucumber and helping to work out levels of fishing that might be sustainable.

At night, the *pepineros* played cards and got drunk. '*¡Muerte al Solitario Jorge!*' came the chant. Inside the CDRS, staff and volunteers, including then director Chantal Blanton and her husband Jim Pinson, took turns to patrol the breeding centre throughout the night, regularly checking on Lonesome George and his fellow animals.

They took the threat seriously. Not only had the recent spate of tortoise killings put everyone on a state of alert, but George had been targeted before. Following a crackdown on sea cucumber fishing back in 1992, *pepineros* from Puerto Villamil on Isabela boated en masse to Santa Cruz, threatening to hold George hostage and set fire to GNPS and CDRS buildings.

That time, before they arrived, head of protection Linda Cayot got wind of their plan; she was not about to put the life of the world's most famous reptile at risk. So Lonesome George was carted out of his corral and a less precious tortoise was put in his place. The animal chosen to act as a stooge for Lonesome George happened to be a female, but there was a similarity, says Cayot. 'Certainly sea cucumber fishermen wouldn't have known that it was not George', she says. As it turned out, the 1992 protest passed without incident.

Within a couple of days, so too did the January 1995 blockade. Troops soon arrived and the protestors slipped quietly away in their boats. Lonesome George had survived again.

More was in store. Following the January uprising, CDRS scientists set about analysing the data collected during the six weeks of experimental fishing. There was also talk of creating a Special Law for the Galápagos, acknowledging that its needs differed from those on mainland Ecuador. Unknown to these parties, Eduardo Velíz, the congressman for the Galápagos, had drafted his own law and succeeded in getting it passed.

Velíz's law changed the definition of a national park to allow 'sustainable use of the natural resources by the local population'. This left room for settlements to expand into protected areas. There were also some serious omissions: no attempt was made to cap the number of Ecuadorians moving to the islands and there was no provision for a quarantine system to stop alien species entering the archipelago. What's more, Velíz's law handed control of the park service over to the politicians. This set a dangerous precedent.

In a letter to her members dated 5 September 1995, Johannah Barry, executive director of the US-based Galápagos conservation charity, the Charles Darwin Foundation Inc., summed things up: 'No country in the world allows treasures of national and international importance to be controlled and managed largely by local political interests.'

On 1 September, the Ecuadorian president at the time, Sixto Duran Ballen, exercised his veto and pulled the new law. When Velíz got wind of this, he began to stir up trouble among fishermen and townspeople. On 4 September 1995, protestors took control of the airports on Baltra and San Cristóbal. They also blocked the entrance to the CDRS and the GNPS compound. 'We are prepared to take extreme steps such as taking tourists hostage if necessary and burning parts of the National Park', Velíz wrote in a letter to the president.

CDRS director Blanton went out to talk to the protestors. She asked that food be allowed through for the staff and animals. They refused, adding that nobody would be let in or out. They would announce their intentions in due course. Things were hotting up for Lonesome George once more.

Around midday, Velíz made an impassioned broadcast on public radio, encouraging the *pepineros* to enter, sack and burn the CDRS and GNPS buildings. Within minutes of his address, a crowd brandishing machetes and bludgeons ran into the compound, baying for GNPS director, Arturo Izurieta. 'They had come to take him by force, they said, then they would kill him', wrote Pinson in one of the regular emails he fired off to friends and family in the US.

The aggressors had heard Izurieta was having lunch in the canteen, but forcing their way in found no sign of their intended victim. He had barricaded himself inside his house along with his two young children. His wife, a natural history guide, was out on a boat in the archipelago, unaware of the drama unfolding within the park service grounds. Unable to get their man, the protestors tore down the gate to the docks and left. Izurieta and his children made their escape along a track at the back of the complex.

That afternoon, word reached those inside that the fishermen planned to set fire to GNPS buildings. By the time darkness fell, the rabble was drunk and the threats of violence increased. Fearing for the safety of her personnel, Blanton decided to evacuate all the students and visiting scientists to a hotel on the other side of the bay.

One of them was Toral-Granda, by now studying sea

cucumbers for a Masters degree. 'We were asked to leave all personal belongings', she remembers. 'We left around midnight in a small boat while the fishermen shouted threats.' Nets had been staked out in the shallow water surrounding the research station to make their escape difficult. Finally, they made it through and out into the darkness of Academy Bay, leaving the chanting hanging in the night air behind them.

Although she'd been advised to get out for her own safety, Blanton stood firm. She and her husband took refuge on an escarpment within the research station, overlooking the iguana-breeding centre. It was raining, a fine misty drizzle. Huddled together, they watched as flames shot up from the barriers in the road and yells bounced off the lava. Every now and then, Blanton's walkie-talkie crackled into action. Until its batteries gave out, she kept in touch with staff in town, carrying out most of the conversations in French. If the protestors were listening in, they probably wouldn't understand, she figured.

In the morning, Blanton made a breakthrough of sorts. At the entrance, she repeated her request for food to be allowed through for the animals and staff. This time, a Guayaquil-based TV reporter captured the whole thing on camera; under the media spotlight, the representative of the fishermen agreed to allow food through the blockade. Normal staff were still prevented from entering the CDRS and GNPS grounds, so the students and visiting scientists – now returned from their evacuation – took up the slack. 'We had to make sure the baby iguanas and tortoises were fine, take care of the aquariums and so on', recalls Toral-Granda.

The following day, things took a sinister turn. Blanton discovered makeshift Molotov cocktails hidden beneath barriers just inside the research station grounds. Back in her office, she took a call from the leader of the protest – a local pharmacist – who threatened imminent action. The renegade

Velíz delivered another inflammatory speech on TV and, getting restless, the protestors hatched a plan to enter the CDRS and take staff and animals hostage. The intensity of the chanting grew. It was a long night, with students taking it in turns to watch over the famous giant. Still nobody made a move on him.

Still under siege a week later, the CDRS and GNPS staff received tragic news. On the morning of 13 September, long-time research station associate Arnaldo Tupiza drove up into the highlands of Isabela to continue work in the *Scalesia* forest. He would normally have been in his jeep, but with it impounded by protesters, Tupiza took a motorbike. On his descent at around noon, he collided head-on with a truck.

Tupiza's death cut through the tensions. The *pepineros* on Santa Cruz relaxed their blockade to let GNPS and CDRS employees attend the funeral on Isabela. On 15 September, hundreds of people followed Tupiza's casket through the streets of Puerto Villamil. Although the Isabela *pepineros* maintained a blockade of the GNPS offices in the town, they did not interfere with the funeral. When the mourners from Santa Cruz returned home, they found the protestors gone. The government had agreed to deal with their complaints. The siege on the institutions in Puerto Ayora had fizzled out.

Officially, the sea cucumber fishery remained closed for the next three years.

Conflicts like this, between conservation aims and the interests of local people, are being played out the world over. It is helpful to divide these up into several different categories. The most difficult to tackle is the straightforward exploitation scenario, where one group of people (usually locals) are removing the flora and fauna around them in an unsustainable manner and another group (often outsiders) would like to stop them. The slaughter of African wildlife for bushmeat and the logging of the Amazon rainforest fall into this category of conflict. For most of the 1990s, so did the exploitation of Galápagos sea cucumbers.

Such conflicts are never as simple as a battle between good and evil. Since 1990, when the Convention on International Trade in Endangered Species (CITES) made it illegal to buy and sell ivory, a lot of attention has focused on the activities of poachers who continue to flaunt international law. Less attention is paid to those who continue to encourage this market. In 2005, a report found ivory products on sale in 776 places in the UK – more than in any other country in Europe. Not only are these retail outlets flaunting the CITES ban, but so too are countless UK citizens who are obviously still set on buying ivory goods. Without this demand, there would be little or no poaching in Africa.

International markets like this and countless others for, say, African bushmeat or Amazonian wood offer local people a quick escape from poverty. They also act as a breeding ground for corruption. Understanding, unpicking and addressing such complex economic, social and frequently dangerous forces is daunting and takes time.

Another set of conflict situations arises where humans and wildlife just don't mix. The Sariska Tiger Reserve in Rhajasthan in India is nearly 900 km² and has more than 300 staff, a good network of roads, 14 vehicles, anti-poaching camps at its remotest corners and a zoning system to clarify what can be done where. It sounds like it has everything going for it. The snag is that its central zone, an area of around 400 km² of prime tiger country, contains 11 villages. For many reasons, spanning many decades, these people have developed a hatred for tigers and reserve managers and have played an active part in poaching out the last animals. In December 2004, Indians were shocked to learn that in spite of all the conservation measures in place in Sariska, it has probably lost its last tiger.

In comparison to the cultural change needed to control commercial exploitation of a species or ecosystem, this kind of conflict is treatable. In the light of the bad news from Sariska, the Indian government's Ministry of Environment and Forests set up a tiger task force to get to the bottom of what went wrong so that other reserves don't suffer the same fate. It's a tall order, but if Sariska's problems can be straightened out, tigers could one day be reintroduced.

Elephants and farmers in Africa enter into similar conflict. Protecting crops by fencing them in or surrounding them with ditches is expensive, time-consuming and doesn't work; the elephants quickly find a way through to the crop or grain stores. Beating drums or letting off guns works – up to a point. It also requires someone to stay up at night and patrol the fields. While a combination of these approaches is typically used to minimize the impact on the life of farmers in sub-Saharan Africa, there is one rather interesting approach that shows the sort of innovative thinking that can help to reduce conflict.

'Elephants hate chilli' is the strapline of the Chilli Pepper Company, an organization that emerged in 2000 in the mid-Zambesi valley of northern Botswana. The idea was to get farmers to switch from crops like cotton that get battered by elephants to fields of chilli peppers, which elephants can't abide. This relieves worry about crops being overrun. With chilli sales on the up, it also brings them a handsome return.

Sometimes, it's not so much cunning as education that's needed. The steady expansion of human settlements into the Ethiopian highlands spelt trouble for the Ethiopian wolf, the world's most endangered member of the dog family. Many villagers saw these carnivores as a threat to their livestock and didn't think twice about picking up a gun and taking aim. Worse, domesticated dogs brought rabies to the wolves. Since 1995, a community-based conservation initiative has reversed these trends. Working closely with villagers, Oxford zoologist

Claudio Sillero has persuaded them to see this rare wolf as the key to a nascent ecotourism industry from which they can benefit. They have become its custodians and are vaccinating their dogs against rabies to reduce the incidence of the killer virus.

Figure 6.2 The Ethiopian wolf is the rarest member of the dog family in the world

For several decades, it has been assumed that this kind of guardian model is one of the best ways to encourage a conservation mindset, giving local people an economic incentive to live in harmony with their surroundings rather than exploiting them unsustainably. There are plenty of examples, like that of the Ethiopian wolf, where it seems to be working.

That said, there is a limit to the amount of ecotourism a particular experience can generate and so too to the size of local population it can sustain. A tourist lodge in Peru – the Infierno Community Ecotourism Project – undoubtedly brings in money for local people. But a study revealed that only one family, whose adult members were all employed by the lodge, earned enough to sit back on the proceeds of ecotourism. Others in the community profited, but nowhere near enough to live on. They had to resort to other money-making activities, many of which undermined the ecotourism operation. Unless the needs of an entire community can be met, the environment will continue to suffer.

The controversial solution, argues Paul Ferraro, an economist at Georgia State University specializing in environmental policy, is to pay local people *not* to damage their surroundings. It's rather like the European Union paying farmers to set aside fields to improve the rural landscape and wildlife.

The logic is simple, Ferraro argues. If you are driving from point A to point B with just one tank of petrol, do you take the direct route or the scenic route? Clearly, opting for the direct route improves the chance that you'll get there, says Ferraro. Using scarce conservation funding to encourage an ecotourism business is the circuitous route. You get what you pay for, so you should pay for what you want to get, he says.

This method of paying locals to act as custodians has been tried in a few places. A long-standing project in Costa Rica aims to protect the country's forests by paying local landowners to lay down their chainsaws. In the second half of the 20th century, forest cover in Costa Rica fell from around 50% to 25%. More than half of what remains is on privately owned land. Since 1996, the government has given individual landowners, associations thereof or indigenous reserves annual payments for conserving forest on their patch. If they don't, they don't see the money. So far, more than 10,000 km^2 of Costa Rican forest has been protected in this way.

●

The unrest in September 1995 did not achieve much for the *pepineros* – the sea cucumber fishery remained closed for several years. Nonetheless, the exploitation continued; it's easy to see why.

Fishermen face the so-called 'tragedy of the commons'. In 1968, microbiologist-turned-ecologist Garrett Hardin published an influential essay in *Science*. He imagined a patch of common pasture on which anyone is free to graze cattle. Farmer A, keen to make the best living he can, adds one more cow to his herd. Then another and another.

Each animal he puts out to pasture makes it just that little bit less lush. This cost is spread between everyone. Farmer B sees what's going on; the pasture is getting worse. Clearly, it would be best for everyone if there were a limit on herd size. Farmer A won't listen and continues to expand. Farmer B doesn't like it, but knows he too must expand. With Farmer A carrying on the way he is, the pasture will soon be wrecked; best make the most of it while there's still grass to be grazed, he reasons.

With some fishermen flagrantly flouting the sea cucumber ban, others surely succumbed to Hardin's logic. The GNPS valiantly tried to keep a lid on this illegal activity. On 4 March 1997, the park service patrol launch *Guadalupe River* approached an illegal camp on the west of Isabela, where wardens found and confiscated 5000 dried sea cucumbers. The next day, on the northwest coast of Fernandina, they took a further 22,000 dead animals. Then, cruising back down the western coast of Isabela, they encountered a Guayaquil-registered boat named *Magdalena*. On board, they found 25 sacks containing around 30,000 dried *pepinos*.

This kind of work was not without dangers. On 19 March, another patrol boat – the *Belle Vie* – motored up the west coast of Isabela. The wardens were looking for an illegal *pepinero* camp clocked during a recent aerial reconnaissance of the area. There it was, clearly visible. Just before 3pm, they began to disembark to inspect the camp, when they were ambushed by around 20 heavily armed men. Shots rang out and a bullet struck Julio Lopez, one of the *Belle Vie*'s crew, in the belly. Fortunately, there was a helicopter nearby that airlifted Lopez to the hospital on San Cristóbal. He underwent two operations. More than 300 people from all walks of Galápagos life came out onto the streets of Puerto Ayora to show their support for the park service. Some carried placards wishing Lopez a swift recovery.

At last, in March 1998, a new Special Law for the Galápagos showed a way out of the sea cucumber crisis. The Galápagos Marine Reserve replaced its troubled predecessor the Galápagos Marine Resources Reserve. This new area was extended to cover an impressive 130,000 km^2 of ocean, protecting underwater sea mounts from industrial-scale fishing and making it the second largest marine reserve in the world after Australia's Great Barrier Reef National Park. Crucially, the Special Law set up a new system of management for the waters, bringing the GNPS, CDRS, fishermen, tourist industry and natural history guides around the table for the first time. This ushered in a new era for conservation in the archipelago – one based firmly on cooperation. The future of the reserve was, and still is, resting on how well these different parties can work together.

The Special Law was an incredible achievement that offered everyone a way to navigate out of conflict. The next challenge was to put it into practice. In 1999, there was agreement over how the sea cucumber crisis should be managed. It would be legal to take the sea-floor bounty for two months each year, as long as the *pepineros* stuck to a quota and helped record the number of sea cucumbers taken and survey populations before and after fishing. Since then, Toral-Granda has been studying the impact of the fishery. She has worked with park wardens and representatives of the fishermen to monitor about 50 different sites around six of the islands.

From 1999 to 2003, more than 25 million sea cucumbers – that's nearly 7000 tonnes in live weight – were legally collected. Unsurprisingly, this is having a serious impact on the sea cucumber population. In 1999, a diver could expect to collect over 35 kg per day. By 2003, 25 kg was hard to come by. Over the same period, the size of an average sea cucumber dropped from 25 cm to 20 cm. Not only were fishermen struggling to find them, but the sea cucumbers were struggling to find each other to reproduce.

What is the alternative? None, says present CDRS director Watkins. Working closely with the fishermen, it has been possible to foster the cooperation essential to make conservation work. 'It allows us to begin discussions about sustainability and about why sea cucumber stocks are declining', he says.

Change was never going to happen overnight. Since the Special Law of 1998, there has been some progress and some setbacks. Fishing continued outside the two-month legal window, with GNPS authorities making regular arrests. On 28 December 2003, following a tip-off, police stormed a building on San Cristóbal. They found 12 parcels, wrapped in plastic, containing just under 16,000 processed cucumbers. In March 2004, port authorities inspecting a cargo ship in Puerto Villamil on Isabela came across bins and a freezer containing nearly 20,000 sea cucumbers. Presumably, these coups accounted for just a tiny proportion of the creatures being caught out of season.

Political instability did not help, with several changes to the Ecuadorian presidency, many different ministers of the environment and a steady stream of directors to the Galápagos National Park Service. When, in September 2004, the government appointed the eighth director in just two years, the wardens reluctantly went on strike. Like the fishermen before them, they set up a barricade at the entrance to the park service offices, this time to prevent the latest incumbent from taking office. Their demands? That the appointment of director should be based on merit rather than politics and fixed for several years to bring much needed stability and leadership to the organization. In 2005, Washington Tapia, a senior warden and respected local, took up office, holding the fort until an acceptable long-term appointment can be made.

Similar tensions exist over exploitation of other creatures in the archipelago, such as lobster. Yet the twists and turns in the tale of the Galápagos sea cucumber, of which this is only a flavour, best illustrate the forces that pull the islands in different directions. Maybe now this unattractive, unassuming species will provide the inspiration for those same forces to pull together – and free Lonesome George from death threats.

Chapter 7
THE MYSTERIES OF PINTA

Decades have passed since the surprise discovery of Lonesome George. There has always been a lingering suspicion that there might be another tortoise, even tortoises, left on Pinta. It's a small island – just 11 km from top to bottom and 7 km from side to side – but its remote location on the northern fringe of the archipelago makes it one of the least explored. Herman Melville's gloomy account of the Galápagos nicely captures the isolation of Pinta, then known as Abingdon or Abington Island:

> Yonder, though, to the E.N.E., I mark a distant dusky ridge. It is Abington Isle, one of the most northerly of the group, so solitary, remote, and blank, it looks like No-Man's Land seen off our northern shore. I doubt whether two human beings ever touched upon that spot. So far as yon Abington Isle is concerned, Adam and his billions of posterity remain uncreated.

Little has changed. Pinta is visited rarely by park wardens, infrequently by scientists and never by tourists. In fact, it's probably safe to say that since Melville wrote his pen portrait, only a few hundred humans have breached the island's rocky defences.

Peter Pritchard has done so twice. We saw how he first set foot on Pinta, back in 1972, the day after Lonesome George was discovered. More than 30 years later he went back with a crack team of park wardens to scour the island for signs of tortoise life. Several members of the group had been before as a part of the decade-long battle to eradicate goats.

Of the alien invaders to reach the Galápagos, large mammals have had by far the greatest impact. When humans came to live on the islands from the mid-19th century onwards, they brought livestock, such as pigs, goats and cattle. Settlers released many of these animals onto uninhabited islands like Pinta to provide food for fishermen far from home.

Some entrepreneurs ran remote islands as farms, coming ashore every now and again to shoot hundreds of animals, then carting them back to sell in mainland Ecuador. In *The Galápagos Affair*, John Treherne recounts that by the 1930s pigs were rampant on Floreana. He tells the story of a German couple, Friedrich and Dore Ritter, menaced by these pigs. 'Friedrich found himself involved in a protracted private war with one most troublesome boar. This enemy he credited with satanic qualities. He first attempted to shoot it with his inadequate rifle; then he tempted it with poisoned bananas.' When this failed, Ritter built a pitfall trap, an elaborate guillotine and a device designed to crush the animal under logs. Finally, he resorted to dynamite. This did not kill the beast, but the shock gave the Ritters and their crops some respite.

The arrival of large mammals was a catastrophe for the Galápagos. Nowhere is this more neatly illustrated than on Santiago, where pigs were introduced some time after Darwin's 1835 visit. They wrought huge changes, preying on plants, invertebrates, lava lizards, green turtles, Galápagos petrels and snaffling up tortoise eggs and hatchlings. In 1968, the GNPS and CDRS launched a pig control programme. After a patchy start, the operation has been carefully coordinated. Unrelenting hunting trips and cunning use of poisoned bait brought the pig population close to its knees; the last animal was dispatched in April 2000.

The story is similar on Pinta. Fishermen released a couple of goats here in the late 1950s. They bred like billy-o, opening up forest and scrub, bringing *Scalesia* and *Opuntia* trees close to collapse and triggering widespread soil erosion. The Pinta tortoise had probably been perilously close to extinction for some time; with the arrival of goats, all hope of finding a female tortoise on the island began to fade. Since the discovery of George, park wardens scoured the island without finding a single living tortoise. In 1996, with the only known specimen of the subspecies in captivity, the World Conservation Union (International Union for the Conservation of Nature and Natural Resources; IUCN) announced that the Pinta tortoise was formally 'Extinct in the Wild'.

Declaring a species extinct is not a step that can be taken lightly. Sounding the knell too early can imperil it further. The Cebu flowerpecker *Dicaeum quadricolor* was a common bird on the Philippine island of Cebu, until large swathes of forest were cleared to make way for agricultural land in the early 20th century. In 1959, the flowerpecker was declared extinct and entered the textbooks as a classic case of a species snuffed out by habitat destruction. This meant that people stopped looking for it. In 1992, birdwatchers spotted some flowerpeckers in a scrap of original forest by chance. The Cebu flowerpecker survives – just. If it hadn't been declared extinct all those years ago, this tenacious population might have been found sooner and more done to protect it.

The recent rediscovery of the ivory-billed woodpecker *Campephilus principalis* is a counter example. Until 2004, nobody had seen this massive North American bird for 60 years. Although most people assumed that it was no more, the IUCN continued to classify it as critically endangered rather than extinct. With hindsight, this was fortunate. The tantalizing prospect of catching sight of a critically endangered species kept a lot of people looking. One, a professor of electronics and computing at Arkansas University, was canoeing down the swampy White River in Arkansas with his video camera when he managed to capture four blurry seconds of something flying through the trees. The footage was enough to convince ornithologists that the ivory-billed woodpecker lives. Had it been declared extinct, it may never have been found.

Could the IUCN be wrong about the Pinta tortoise? Might one or maybe more of George's amigos still be roaming the most inaccessible parts of his wild home? There are several reasons to think so.

Pinta once had a thriving population of tortoises. The logbooks of 19th-century whaling vessels suggest that the main reason to stop off at the island was to collect the animals for food. William Ambrosia Cowley, an English-born buccaneer who hung out in the archipelago in 1684, penned an early description of the place: 'We, with 7 of our men, went to discover this island, to see if we could get any water or turtle there', he told his journal (by 'turtle' he meant tortoise). 'We came with the boat to the island, but found it so steep that we could not get upon it. We sailed along the east side of it, thinking to get a place there, but it growing towards night, and seeing our ship a great way from the land, we stood away to the northernmost part of the island to meet her.' When Cowley and his men finally made it onto Pinta, they found it an inhospitable place. 'The land had been burnt formerly, and the rocks were split in pieces by some sulphurous matter that had taken fire.' Cowley did not find tortoises, although he suspected they were there.

The discovery of the Pinta tortoise did not occur until 1798, when a certain Captain James Colnett moored alongside the island (which he refers to as Abington). He wrote:

We were close under Abington Isle, which is very small, and was well known to the Buccaneers ... It is high towards the South end, which has a very pleasant appearance, and where is the only

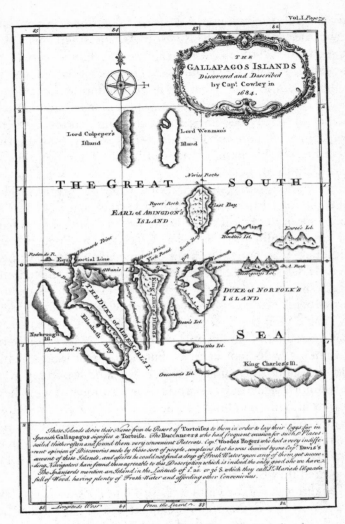

Figure 7.1 The Galápagos Islands, as discovered and described by Captain Cowley in 1684

bay or anchoring place in the island. The North end is low, barren, and one entire clinker, with breakers stretching out to a considerable distance. I sent a party in the boat to round it, where they caught plenty of small fish with their hook and line. They also landed on the island and found both tortoises and turtles.

From then onwards, the logbooks read like a series of efficient and devastating raids. In the late 19th and early 20th century, the emphasis shifted from eating tortoises to taking them for scientific collections. By then, the Pinta race was so scarce that several research expeditions failed to get any specimens at all.

The discovery of George in 1971 means they overlooked at least one. There were probably more. In 1964, CDRS employees Miguel Castro and David Cavagñaro found the remains of 28 tortoises trapped at the bottom of ravines, most likely part of a small population that survived well into the 20th century. Did any tortoises, apart from George, make it into the 1970s and beyond? Almost certainly.

In 1976, just four years after Lonesome George had sailed off to Santa Cruz, a small research team landed on Pinta for a week-long expedition to study mosses and lichens. There were three botanists – Rob Gradstein, Stig Jeppesen and Harrie Sipman; assisting them was a park warden and a young CDRS zoologist Derek Green.

Gradstein, now head of systematic botany at the University of Göttingen in Germany, remembers this expedition as one of the most thrilling of his Galápagos trip. 'Pinta at the time seemed to us the most unspoiled of all the islands in the archipelago – some kind of Eden – visited by scientists on average only once a year', he says. It took nearly two days on board a smelly fishing boat for the team to reach the southern tip. In the late afternoon, Gradstein and his colleagues pitched camp and cooked dinner. 'While preparing our meal at the camp, finches and lizards would come near our plates', recalls Gradstein. 'The Galápagos hawk would fly overhead ready to snap the fried meat out of our pan.'

On 9 July, the men set out across the lava field at the southern end of the volcano. They planned to walk up to the summit and back. That way, they would cover all altitudes on the island and be able to record the different mosses and lichens that thrive at different heights. Back then, virtually nothing was known about Galápagos mosses and liverworts (collectively, the bryophytes) and lichens. 'I don't think there were any species records, so this was very exciting, kind of exploring a white space on the map', says Gradstein.

Over four days on Pinta, the researchers recorded about 70 species of bryophyte – more than a third of all those now known in the archipelago. For such a small island, this is remarkable. Part of the reason for this richness, suggests Gradstein, is that Pinta is one of the higher volcanoes, so offers a spectrum of habitats that suit different species. There is an arid zone at the base, humidity increases further up and moist exposed rocks sit near the top.

Although Derek Green was keen to learn about plants and fungi, he did not quite share the botanists' exhilaration as they unearthed one new species after another. Whatever way you look at it, he says, moss is still moss. Green had spent a year in Surinam catching, tagging and releasing green and hawksbill turtles. He had come to the Galápagos to work as a guide on one of the tour boats before taking up a position at the CDRS. This Pinta trip was one of his first jobs for the research station.

While the park warden hacked a path through the under-growth with a machete, it was Green's task to give the botanists scientific back-up. With his all-round knowledge of the archipelago's flora and fauna, he was well placed to make observations of interest to other scientists who might visit in years to come. When the botanists came to a halt halfway up the mountain and began to dig around for mosses and lichens once more, Green took the opportunity to wander off to explore. It was his 27th birthday. 'On that particular day I just had this gut feeling that I was going to find a tortoise', he says. Sure enough, he soon came across the remains of a large one wedged in a crevice.

The bony shell of a live tortoise has a layer of protective plates called scutes. When a tortoise dies, these begin to fall off. Once the bony shell is exposed to the harsh equatorial conditions, it disintegrates pretty quickly. The shell of the Pinta tortoise is thinner than most, so crumbles in months. Within a couple of years, all that's left is a small heap of weathering bones.

Green's tortoise was still in pretty good shape. The shell had three scutes still in place. With help from Jeppeson, Green climbed down the steep lava wall, freed the shell and dragged it up out of the crevice. He then dropped to the bottom of the ravine to bag up the rest of the scutes and bones that had tumbled out. The botanists were sufficiently impressed to take a break and pose for a photograph with the empty shell.

They left the find to pick up on their descent. Once they reached the summit, did a bit more collecting and dropped back down, they lashed the shell to the frame of a backpack and Gradstein posed for a photograph. It is now at the CDRS, where Peter Pritchard has been able to study it in detail. 'It hadn't been dead more than a year or so', he says, 'maybe less.' If Pritchard is right, Green's birthday discovery suggests that there was still at least one tortoise alive on Pinta after Lonesome George was removed. The size of the shell and bones indicates that this animal was also male.

Figure 7.2 Botanist Rob Gradstein poses with the
intact shell of a Pinta tortoise on 9 July 1976

On 26 March 1981 – nearly five years after the moss-mapping
trip and almost ten years since Lonesome George last trod his
home soil – Linda Cayot made another discovery. She was not
meant to be on Pinta that day but on Santiago to begin her
own research into the foraging behaviour of giant tortoises.
But her field assistant was ill. With nobody to accompany her
as GNPS regulations require, Cayot joined botanist Ole
Hamann and his young student Ole Seberg on nearby Pinta.

The discovery, which she almost stepped on, was what is
technically known as a scat. In other words, Cayot found a poo.

This was Hamann's second trip to Pinta. On his first, 10 years earlier – when Lonesome George was discovered – he had begun a long-term project to assess how the island's vegetation was responding to the goat cull. He had marked out several small square plots at a range of altitudes to monitor the changing vegetation. It was about halfway up the volcano that Cayot found the scat, in an open area dotted with umbrella-like *Scalesia* trees.

'I can still envision the precise point', recalls Cayot. 'I was following the two Oles across the slope above the saddle of the island, and as I stepped across a grassy area I looked down', she says. 'I yelled out and we all stopped.' Hamann and Seberg had stepped right over it, perhaps because, like snail biologist Vagvolgyi before them, it was not what they were looking for. They heard Cayot's cry, turned and came running back to where she was standing. 'We were all amazed', she says.

They all remember the poo: 'It was a single small scat and was very dry', says Cayot. 'Definitely sort of old and withered', is Hamann's version. Seberg is more poetic: 'Greyish-brown, with a coarse, fibrous structure, elongated cigar-like but with a circumference of a tennis ball.' There is no doubt that it came from a giant tortoise.

The Pinta poop occupied Cayot's thoughts for the rest of the trip. Its crusty condition meant it was not a recent deposit. But it had been lying on a patch of grass out in the open, where it would have been battered by the elements. Could a George dropping have really lasted a decade in such an exposed spot?

In a normal year, the trade winds blow strongly across the Pacific from South America to Southeast Asia, driving the waves from east to west. Cold, nutrient-rich water is drawn up from the depths of the ocean off the coast of South America. In an El Niño year, this sequence breaks down: the trade winds slacken, the upwelling of cold water slows, the surface water and the air begin to warm. The upshot is rain – a lot of it. The wettest year since records began was in 1982, when the islands were awash with about 10 times the normal rainfall.

In 1975 and 1976, another strong El Niño whipped up ferocious weather, which would almost certainly have obliterated any signs of a scat on a bare mountain. This implies that the Pinta poo was not George's. Even Green's 1976 tortoise probably died before the El Niño. This reasoning implies that a third tortoise defecated some time between 1976 and 1981.

If DNA could be pulled from the Pinta poo, it would, in theory, be possible to work out if it had come from Lonesome George or another tortoise. Unfortunately, it's unlikely that much DNA remains after all this time. It sat on a tuft of grass beneath the equatorial sunshine for several years, was carried to the CDRS where it became a museum exhibit for a few more and was finally packaged up and stored in a box in some backroom in the research station. These are not ideal conditions for DNA preservation.

It's surprising just how much interesting information faeces can reveal. The Iberian lynx *Lynx pardinus* is the most threatened cat on the planet. So elusive is this solitary and nocturnal creature that estimates of population size are difficult to come by. Its last stronghold is thought to be the Doñana National Park in southwest Andalusia in Spain. There are probably a few other places where one or two individuals cling on. These animals are important for any conservation effort because their genes are likely to be a bit different from those in Doñana, and when populations get small, the more genetic diversity the better. In the late 1990s, biologists identified five places in Portugal that might still harbour small lynx populations.

Concrete evidence, such as sightings or the odd dead animal, was worryingly sparse. Fortunately, 104 scats were collected from various sites in the late 1990s that might have come from lynx. This gave an opportunity to estimate how many Portuguese animals there might be. There are two problems with just counting scats. First, one can't be certain that a scat actually belongs to a lynx. 'It is possible to misidentify scats and, for instance, confuse lynx with wildcat scats', says Margarida Fernandes of Portugal's Instituto da Conservação da Natureza. Second, 104 scats does not, of course, mean 104 lynx. It could mean just one rather regular lynx (although given that the five sites cover the length of the country, this is, admittedly, unlikely).

To bring a little clarity to the situation, Fernandes looked to DNA analysis. When an animal defecates, cells from its intestinal wall are jettisoned along with the faeces. There is a branch of biology, molecular scatology, dedicated to the scrutiny of such cells (and other microscopic faecal matter). In the case of the 104 scats, DNA analysis revealed that only two were from lynx. This suggests that the populations identified in the 1990s have disappeared – bad news for the future of this endangered cat.

In the remote Yellowhead Ecosystem in Alberta, Canada, faeces are influencing the management of bears. Finding bear scats over an area of more than 5000 km^2 is no easy task. Canadian ecologists have trained dogs to search out bear poo – they are about five times better at finding it than are humans, says Samuel Wasser, director of the Center for Conservation Biology at the University of Washington. With canine help, Wasser collected 840 bear scats between 1999 and 2001. These could have come from grizzlies or black bears. DNA revealed which dropping belonged to which species, which individual bear was responsible and its sex. The two species seem to frequent different parts of the Yellowhead Ecosystem and for grizzlies, it looks like there are nearly two females to every male.

Figure 7.3 The number of grizzly bears in an area
can be estimated by counting faeces

Alternatively, faeces can offer a way of collecting DNA
from animals that are tricky to handle. Kim Parsons, of the
National Oceanic and Atmospheric Administration, is after
dolphin DNA. Snorkel-clad scientists tread water while
waiting for a dolphin to defecate. Then they pounce,
gathering up sinking faeces in small plastic vials. A similar
technique could be used for other aquatic creatures, even
fish; but the smaller the species and the closer it is to the
bottom of the sea, the harder it is to collect samples, warns
Parsons. 'You have to be able to locate, swim to and collect
the faeces before they land on the seafloor.'

Green's tortoise and Cayot's scat make it entirely plausible
that there might still be a tortoise on Pinta. Giant tortoises,

particularly young ones, can be surprisingly hard to spot, says reptile expert Howard Snell, former director of science at the CDRS. Two tortoises were introduced some years ago onto the tiny island of Santa Fe, which is visited by at least 50,000 tourists a year and serviced by three research teams. In spite of this human activity – far more than on Pinta – the duo is seen only every seven or eight years, Snell says.

At a 1998 conference entitled 'Science for Conservation in Galápagos', there was discussion about the status of the mysterious Fernandina tortoise, appropriately named Geochelone nigra phantastica. Only one specimen has ever been found on this island, the youngest, most pristine and least explored in the archipelago, nearly 10 times larger than Pinta.

During the California Academy expedition at the beginning of the 20th century, its leader Rollo Beck spent a couple of days on Fernandina while the rest of the party probed the wilds of Isabela. On the afternoon of 4 April 1906 he came across a tortoise trail, probably used within the last day or so. Even though he'd tracked hundreds of giant tortoises, this one was different. Beck knew that if there was an animal at the end of the trail, it would be one completely unknown to the scientific world. Soon he came across a rock in the middle of the trail and knew he was onto a male. 'It had been used for the same purpose that rocks … have served ever since the whalers carried off all the female tortoises', he later quipped to his friends on board the expedition vessel the Academy.

He soon found his quarry, chomping grass. As the only tortoise ever found on Fernandina, this old male might have become an icon like Lonesome George. But contracted to collect specimens for the California Academy museum and not living animals, workaholic Beck quickly ruled that out: 'Getting my pack, I ate supper and skinned the tortoise by moonlight.'

In 1964, an expedition discovered tortoise-like droppings, which stirred up some interest, although there was a suspicion that these had been dumped by large iguanas. When Himalayan mountaineering legend Eric Shipton scoured the island a few years later, he concluded that the odds of locating a Fernandina tortoise were lower than those of finding a Yeti. Only a handful of delegates at the 1998 conference felt a population might still survive on the remote southern slope where there's sufficient plant life. The Fernandina subspecies, if there ever really was one, probably struggled to survive on an island tormented by frequent eruptions.

It was more plausible to imagine that Pinta might still harbour a few tortoises. 'Using experienced tortoise searchers, and potentially even trained dogs, every effort should be made to investigate the possible existence of additional individuals as potential mates for the single known male', concluded zoologist Thomas Fritts and several key scientists and wardens at the conference.

In 2000, a television company got Fritts as their expert in a show subsequently aired on the Discovery Channel. They chartered a helicopter to fly low over Pinta with high-tech imaging equipment. They found no evidence of tortoise life.

In 2002, Peter Pritchard hatched a plan to return to Pinta to find a mate for Lonesome George. He got in touch with the GNPS. It was a win–win situation. Pritchard would secure enough cash to buy some equipment and the park service would give him permission and a team of rangers to scour the island. Pritchard wrote to the Walt Disney Company asking for $18,000 – enough, he reckoned, to fund the trip. Disney came back with $2000. It was a start. Within a year, local fund-raising efforts on behalf of Pritchard's Chelonian Research Institute had made up the shortfall and the expedition, pencilled in for October 2003, got the go-ahead.

Figure 7.4 Peter Pritchard (back left) and the wardens who scoured Pinta for tortoise life in October 2003

Before travelling to Pinta, the search team went on a couple of mini-expeditions to practise spotting tortoises. The first was a one-day excursion to the little-studied Cerro Fatal population on the rocky eastern side of Santa Cruz's volcano. They found eleven animals that day – eight youngsters, two adult males and one adult female.

The second training trip was to Wolf volcano on Isabela, a good place to break the long journey from Santa Cruz to Pinta. Over three days, the team found 30 tortoises. Surprisingly, only one was already marked, even though a similar expedition had covered the same ground just two years earlier and marked nearly 50 tortoises. At last it was time to begin the search proper, and the boat set out for Pinta.

The search was extremely thorough and included the upland and lowland areas of the island. It did not cover the northerly part or the eastern coast, which are too sparsely vegetated to sustain tortoises. Initially, a line of wardens patrolled one area at a time, walking about 10–25 m apart. They were of course on the lookout for tortoises, but also for droppings, trails through the undergrowth and flattened grass where a tortoise might have slept.

Figure 7.5 Peter Pritchard on the southern tip of Pinta with tortoise bones found on the expedition

Soon team members began to come across the remains of tortoises at the bottom of ravines or trapped in crevices, as Green had done more than 25 years earlier. They conveyed global positioning satellite (GPS) locations of these sites to Pritchard, who followed up to photograph and inventory the bones, and carefully extract what data he could. In most cases, only part of a skeleton remained; usually the shell and underbelly had disintegrated, leaving just a pile of bony rubble. Putting together the pieces, Pritchard was able to identify at least 15 different specimens and work out a lot about them.

The pelvis of the giant tortoise is made up of six separate bones; its shoulder piece comprises four more – two on each side. While an animal is growing and even in early maturity, the joins between these bones are still visible. Some time after the animal has stopped growing, they fuse and the pelvis and shoulder piece each look like a single unit. This allowed Pritchard to infer the age of each specimen. 'Most appeared to be of adult animals and only two were young', he wrote in his report to the CDRS after the trip.

Specimen 13 was in excellent condition. It was found on 26 October towards the end of the expedition nestling in a small pit at the bottom of a deep, wide fissure. 'It was intact, except for a large hole in the left side of the carapace, perhaps made by a falling stone or rock', Pritchard noted. Its fused pelvis and concave underbelly suggested it was an adult male. 'Judging by the condition of the bones, this specimen, at the time of its demise, may have been the last tortoise on the island', he wrote.

All but one of the tortoises seems to have been male. The lone female, specimen 14, was discovered on 25 October southwest of the main peak about a third of the way up. All that remained was a pelvis and one of its limb bones. The pelvis was completely fused, so the animal had finished growing, but it was much smaller than the male pelvises collected on the trip. 'From that, I was assuming that was an old female', says Pritchard. Perhaps this was the last Pinta female.

The expedition found no live tortoises and, more importantly, no signs of any. 'There are areas ... that appear eminently suitable for tortoise nesting but alas show no sign of the past or recent presence of tortoises', lamented Pritchard.

The excess of males is puzzling. There are a couple of

possible explanations. First, females – being lighter and easier to carry – were harder hit by whalers. An entry in Captain Porter's *Journal* confirms this. Arriving on Santiago in 1813, it took just four days for his men to collect over 12 tonnes-worth of tortoise; all but three of these were female. 'They come down from the mountains for the express purpose of laying; and this opinion seems strengthened from the circumstance of there being no male tortoises among them, the few we found having been taken a considerable distance up the mountain.'

Second, reptiles are particularly susceptible to wild fluctuations in their sex ratio, more than say mammals or birds. This is because in most – tortoises included – gender is determined not by genes but by temperature. When a female giant tortoise marches onto the nesting grounds, digs out a shallow pit and lays her clutch of 10 or so eggs, the nascent tortoise inside each egg is neither male nor female: its gender is decided entirely by the temperature of the soil in which it is buried.

Why such a system evolved is still a mystery, but many biologists believe that before mammals or birds appeared on the scene, this form of sex determination was standard for land vertebrates. The most likely explanation is that the temperature during incubation affects the sexes in different ways. If, for some reason, males do best when incubated at one temperature while females do best when incubated at another temperature, then it would be a good strategy if sex could be influenced by environment rather than genes.

In addition to most turtles and tortoises, temperature dictates the gender of many crocodiles, alligators, lizards and skinks. There is some variation in how it works. For some species, high or low incubation temperatures produce females, while an intermediate incubation temperature produces males. For others, cooler conditions elicit females and warmer soil gives rise to males. For yet others, this pattern is reversed.

The Galápagos giant tortoise falls into this last group, so the rule is 'hot chicks' and 'cool dudes'.

It's just possible that some special property of Pinta soil makes it more likely to nurture predominantly male clutches. Of the Galápagos Islands, Pinta does have a very unusual makeup, says William White, professor of geological sciences at Cornell University. Most lava contains the mineral plagioclase. On Pinta, plagioclase is particularly abundant and its crystals unusually large. This, White says, could make the soil cooler than on other islands, which would favour males.

Alternatively, once populations become really small, normal patterns – like equal numbers of males and females – start to break down. Female green turtles return, year after year, to lay their eggs on the beach they hatched out on, often within metres of their natal nest. Less work has been done on the nesting patterns of tortoises, but it is likely that female tortoises will be similarly faithful to a particular nesting ground and perhaps even a particular place within that nesting ground. So a female tortoise with a preference for nesting in a sunny spot will, clutch after clutch, produce more females. In a large population, such individual preferences even out. Once the population has dropped to just a few animals, these idiosyncrasies become apparent. In the most extreme case, if the last remaining female – possibly specimen 14 – had a penchant for laying her eggs in the shadow of a favourite cactus, then most and perhaps all of her offspring would have been males. The whalers may have taken most of the female tortoises from Pinta and perhaps those that remained produced exclusively male clutches.

Although it cannot be ruled out that the 2003 expedition overlooked a tortoise on the unpopulated and overgrown island, it is now time to look for a Pinta female elsewhere. The search continues in the world's zoos.

Chapter 8
THE DIASPORA

There are probably about a thousand Galápagos giant tortoises in zoos and private collections around the world. It's a bit of a long shot, but it's just possible that one of them is a Pinta animal collected from the island in the late 19th or early 20th century. Ever since the Charles Darwin Foundation offered a reward of $10,000 for a Pinta female, the search has been on in this large population of captive tortoises.

Most of the tortoises taken away in ships down the ages didn't make it far. They were just too tasty. Then, at the end of the 19th century, along came science.

With so few Galápagos tortoises left, the race was on among naturalists to scour the islands for the last ones. 'There was a sense that this opportunity was going to end', says Edward J. Larson, author of *Evolution's Workshop*. During the 1880s and 90s, a fabulous accumulation of industrial wealth led to a surge in private money being pumped into museums, universities and zoos. At the same time, there was an increase in popular interest in exotic creatures, scientific interest in evolution and significant advances in sailing technology, says Larson. All these conspired to create a tortoise rush.

This collecting frenzy that peaked at the turn of the century means that you don't have to go to the Galápagos to see a giant tortoise. Every respectable natural history museum got its hands on one or two and the best nabbed hundreds. Zoos were also keen to acquire them, as these extraordinary beasts brought more visitors through the turn-stiles. Might one of these zoo-based tortoises be an ageing female from Pinta to put the spark back in Lonesome George's love life?

It's a good question and surprisingly difficult to answer. The biggest problem is that nobody can be quite sure what a Pinta female actually looks like. One sensible step would be to find one in a museum and mug up on its size and shape. Unfortunately, by the time people had given up eating and started collecting tortoises, the Pinta population was already thin on the volcanic ground. So in spite of all the efforts, we know of only a handful of tortoises that came from Pinta and almost all of them are male.

Colin McCarthy, curator of reptiles at London's Natural History Museum, leads the way into one of the storerooms that house the 'type specimens' in the collection – the specimens used to describe a species new to science. These precious individuals – often centuries old – are stored in glass cabinets on shelf above shelf, in corridor after corridor, huddled together out of the public gaze as a physical reference point for subsequent discovery.

Figure 8.1 Colin McCarthy, curator of reptiles at London's Natural History Museum, with the type specimens of the Pinta tortoise

In one corner of this dusty vault, just beneath the eager stampede that daily pounds the museum's entrance hall, are the type tortoises. There are tiny tortoises at the feet of medium-sized tortoises among a handful of giant tortoises. Some of the giants come from islands in the Indian Ocean and some of them from the Galápagos. Three of the Galápagos tortoises, collected in 1875, are the type specimens of the Pinta subspecies – the ancestors of Lonesome George that brought his clan to the attention of the scientific world. These tortoises were some of the first acquisitions in the collecting frenzy that lasted some 30 years. Their discovery guaranteed them quite extraordinary lives.

＊

Commander W. E. Cookson swung a leg out of the skiff and onto a ledge of black lava on Pinta. He was determined to collect a giant tortoise from the island and bring it back to England alive. This creature, Cookson calculated, was of serious scientific interest and would make a great attraction at London Zoo. Back on board HMS *Peterel*, moored just off the island, there were already around 30 tortoises. These had come from Isabela, from the same spot that Darwin had visited 40 years earlier.

'They are still tolerably numerous near Tagus Cove', Cookson wrote to his boss, the First Lord of the Admiralty. In addition to these Isabela tortoises, Cookson wanted some from another island. According to local inhabitants, Pinta was his best bet. One of these informants, an Englishman based on Floreana who had been making tortoise oil for 15 years, agreed to help scour the island. 'Without his guidance we should, I think, have failed in our search', wrote Cookson in May 1876 of his Pinta experience.

Even with local knowledge, it took about 25 men two days of sweating graft to find just four tortoises. 'One of these, owing to want of sufficient hands to carry them all, was killed on shore. The other three we brought on board the ship', wrote Cookson.

They had to be carried, slung from poles, a distance of six miles through a bush so thick that a trail had to be cut for the whole distance; besides which the whole surface of the island is covered with irregular blocks of lava, making the walking with a heavy load exceedingly tedious and laborious.

Figure 8.2 One of the Abingdon tortoises collected by Cookson

So determined was Cookson to bring these animals back alive that he was even prepared 'to lower them over the cliff, a height of about 200 feet'. This whole experience must have been something of a shock to the tortoises. One day they were sitting happily amid stunted bush and high, coarse grass; the next they were being inched down a sheer cliff face and into HMS *Peterel*. 'During this process, and more especially in the carrying-down, they received some rough treatment', wrote Cookson. 'There was no external injury; but whether caused by this or not, these tortoises never thrived.'

Within two months, the three Pinta creatures were dead. Cookson decided to sacrifice the smallest of the animals almost immediately. 'I could not preserve all alive, and I hoped to keep the two larger ones', he explained. But these didn't last long either. One was too big to preserve but Cookson skinned its head and feet and stored the dried remains, together with some bones, inside the upturned shell. The other tortoise was a little more fortunate, but not much. It survived until they reached Honolulu. There, Cookson managed to get hold of some spirits from HMS *Challenger* and sent the pickled tortoise to England on board another ship HMS *Repulse*. After the lengths to which he had gone to get these tortoises, their death came as a bit of a blow: 'I greatly regret that the two Abingdon [Pinta] tortoises, which I hoped to have kept alive, died after being a few weeks on board.'

Just over 30 years later, the California Academy of Sciences spent a year and a day in the archipelago collecting flora and fauna from many islands. If Cookson's visit was one of the first to collect Pinta tortoises for science, the California Academy expedition was the last.

On 18 September 1906, nearing the end of their stint among the islands, the *Academy* anchored just off the south of the island and the following morning the scientists went ashore. About 200 m above sea level, the undergrowth started to thicken. A wet mist clung to the lifeless peak. 'It is capital country for tortoises', wrote the expedition's reptile expert Joseph Slevin in his journal.

Soon the naturalists came across signs of a tortoise – a fresh tunnel boring into the undergrowth ahead of them. They followed it for some distance before the vegetation thinned out to reveal a small clearing. There facing them, standing atop a large, low rock, was a 'very fat male'. It was drinking from small pools of water; as they approached it raised its long neck towards them. They skinned it and carried it back to the boat. Over the next couple of days, the California Academy explorers returned to the mountain and, following other trails, tracked down two more males.

When they finally left the archipelago a few days later, these three individuals were among 264 tortoise specimens shipped back to San Francisco. They are the same animals the geneticists studied nearly a century later to double-check that Lonesome George had Pinta-type genes.

In the years between these two expeditions another player entered the picture. The eccentric Lord Walter Rothschild was a wealthy philanthropist and amateur naturalist who bought up live giant tortoises to adorn his country estate at Tring in the south of England. He funded his first collecting trip to the Galápagos in 1897. Another in 1901 yielded what appears to be the only complete Pinta female in the world.

Rothschild knew that several types of Galápagos tortoise, including the Pinta one, were close to extinction. He had read Cookson's 1875 letter from the islands warning that gangs of men were working each island, stripping out the lichen to make the purple dye orchil and living off tortoise meat. During Cookson's stay, these orchil pickers had just finished a four-month stint on Española during which they were said to have killed 70-odd tortoises. This did not bode well for the Pinta animals, which might be next. 'I believe they are doomed to destruction directly the orchilla-pickers are placed on the island', wrote Cookson. It's not clear whether the gang ever made it to Pinta. But it's eminently possible.

So in 1897, Rothschild funded an expedition organized by American businessman Frank Webster and led by naturalist Charles Harris. Rothschild instructed Harris to bring away every tortoise they saw, 'big or little, alive or dead ... to save them for science'. He did not yet have a Pinta tortoise in his collection so included it on his wish list of species.

Just before lunch on 12 August 1897, Harris arrived off the west of Pinta. The following morning he took his men ashore in a small boat. 'I instructed each man to collect about twenty birds, and be back at the boat at 11', Harris wrote in his diary. It would have been a successful morning except that, at the allotted time, one of the team, Frank Drowne, failed to show up. They eventually found him in a muddle on the other side of the island, possibly feeling the effects of dehydration. 'He had got lost, and lost his head', wrote Harris.

Drowne's escapade had slowed things down and they continued to collect on Pinta for a further five days. They didn't find a single tortoise. On 18 August, they set sail for the next island on their itinerary. 'Do not think tortoise exist here; we could find no signs', lamented Harris.

Although the expedition reaped great rewards for Tring's expanding zoological collection, it also highlighted gaps that needed to be filled. Rothschild couldn't tolerate gaps. His tortoise collection, for example, wouldn't be complete without a Pinta specimen.

So in 1901, he contracted the Californian collector Rollo Beck to find examples of species he was missing. Beck had been on the Webster–Harris expedition, so knew the territory. He was also a highly skilled naturalist. This time, Beck managed to track down two tortoises on Pinta. One of these was a large male. The other was a smaller tortoise that could be the only example of a Pinta female that we know of.

The large tortoise was already ancient. Beck killed and skinned this 'old mossback' there and then to simplify the return trek across Pinta's treacherous lava fields. Back on board, he strung out the skin of the unfortunate male to dry. It added to the growing collection of tortoise skins scalped from other islands that Beck shipped to Rothschild that summer.

Beck kept the smaller tortoise alive. He found the animal trapped 'at the foot of a cliff near [the] ocean'. When he and his men clambered down the rock face, they found it had 'but one eye!' Beck guessed it was a female and that she had picked up her injury tumbling down the cliff. If he was right, this small Pinta female must have been trapped beside the ocean for several months. 'The eye is scaled over and scar healed', he wrote in a letter to Ernest Hartert, the curator of Rothschild's private museum at Tring. The one-eyed beast was small enough – around 20 kg – to be lugged to the boat alive where she joined three other live tortoises in a makeshift pen, one from Santa Cruz and two from Tagus Cove on Isabela.

One of the Tagus animals, an aggressive male weighing more than twice that of the Pinta female, did not take too well to its berth. 'One day our largest tortoise got on a "rampage"', Beck informed Hartert. The male extended its neck and tore down several tortoise skins that were drying on a shelf above the pen, taking '2 or 3 bites from legs + heads of the skins'. It may have been in the ensuing chaos that the Pinta female sustained a 'sore leg where bitten by another on board'.

Despite being malnourished after her stint at the bottom of the cliff, having only one eye and sporting a leg injury, the Pinta female survived the trip to California. 'If you wish these as skins I will make them up & send on as soon as dry enough', Beck wrote to Hartert in early July 1901. 'Will wait till I hear from you before killing any.'

It took two months for Rothschild's reply to reach Beck. Keen to add to his collection of living giants at Tring, Rothschild wanted them alive. So Beck set about sourcing a shipping company prepared to look after the tortoises on the trip to England. 'If they will I will ship alive. If not will skin', he wrote on 10 September.

The company Beck approached agreed to ferry the tortoises

to England, but could not guarantee that they would arrive alive. If one of them died in transit, there would be nobody on board to skin and preserve it. Beck feared that it might just be thrown overboard should it die en route. Surely Rothschild would rather have the tortoises dead than not at all, he reasoned. So on 9 November, Beck skinned three of the four tortoises, starting with the Pinta female.

When Rothschild died in 1937, he left his entire collection to the Natural History Museum. Some of the specimens were taken to London; many remained in Tring, where they are now on display in the Walter Rothschild Zoological Museum, an offshoot of the Natural History Museum. Indeed, you can see Beck's 'old mossback' tortoise from Pinta behind glass in the public gallery at Tring. The one-eyed tortoise found her way to the outskirts of London and is in an inaccessible wing of the Natural History Museum's overflow storeroom in Wandsworth.

Figure 8.3 The one-eyed Pinta female
collected by Rollo Beck in July 1901

This specimen is about the sort of size and shape one might expect of a Pinta female. In every other population of giant tortoise we know of, males are at least twice the size of females. What's more, males of saddlebacked subspecies (like those on Pinta) have a more exaggerated furl to the front edge of their shell than females. This general rule suggests that a Pinta female is going to be about half the size of Lonesome George and have a slightly simpler shell.

Beck's one-eyed tortoise fits this bill. Since she may be the only example of a Pinta female in the world, a careful inspection of her size and shape could give some vital clues of what to look for in the search for a mate for George.

—

Museum specimens like the one-eyed female can help solve all sorts of natural history mysteries. Take the case of the supposedly extinct forest owlet *Heteroglaux blewitti*. The owlet is known from just seven specimens, the most recent apparently collected in 1914 by Colonel Richard Meinertzhagen, a British spy who had worked in eastern India. Curiously, nobody could find the owlet in Gujarat, where Meinertzhagen claimed he had bagged it.

Then in the mid-1990s, a close inspection of Meinertzhagen's owlet produced a vital clue. Pamela Rasmussen, an ornithologist working at the Smithsonian Institution in Washington DC, had grown suspicious about the authenticity of many of the specimens collected by Meinertzhagen. Turning the owlet over in her hands, she noticed that the original stitching and stuffing had been removed and that the bird had been stuffed afresh and resewn.

Rasmussen called upon a Federal Bureau of Investigation laboratory to analyse traces of the original stuffing that Meinertzhagen had not managed to remove. It turned out to be a perfect match for cotton used by one James Davidson, a

British official who had been stationed in western India in the 1880s. This and other evidence strongly suggested that this specimen had been collected and stuffed by Davidson and subsequently stolen and restuffed by the unscrupulous colonel.

So, reasoned Rasmussen, the owlet had come not from the east but from the west where Davidson had been. Armed with this knowledge, she and her colleagues tracked down a living owlet perched in a tree near Bombay. 'My first sight of a living, breathing forest owlet quizzically staring down at us was by far the most exciting event of my career', she says. The Bombay Natural History Society is now studying the few known tiny clusters of this species to establish whether there is a viable population that can be saved.

—

The desiccated one-eyed Pinta female, on her own, is not a lot to go on. The appearance of a fully grown tortoise is determined largely by its genes but is also influenced by other factors, says reptile buff Thomas Fritts. 'Periods of malnutrition or rapid growth can sometimes change the shape', he says. Beck's one-eyed find might be typical of a Pinta female. Then again, she might not. Without a whole bunch of other females to compare her to, it's impossible to tell.

There are a couple of other possible Pinta females. The remains of the butchered tortoise that Peter Pritchard found on Pinta just days after Lonesome George's discovery is certainly about the right size. Unfortunately, whoever killed it seems to have taken away its underbelly and bones, which are useful for confirming tortoise gender. Furthermore, the shell is rather an unusual shape, says Pritchard. There is also a stuffed tortoise at the Smithsonian Institution in Washington DC that looks like it could be a Pinta female, says Fritts. But there's no concrete evidence linking this individual to the island.

All this uncertainty over the origin and sex of the museum

specimens means that DNA profiling of zoo-based tortoises is a more reliable way to identify a Pinta female. As geneticists continue to build up the tortoise DNA database, they are finding more and more 'genetic markers' – signature sequences that are unique to a particular species, subspecies or even population.

Not far from Lonesome George's enclosure at the CDRS, there are 23 tortoises confiscated from or handed over by private collectors. Because nobody is sure which subspecies they belong to, they are kept isolated from other captive tortoises of known pedigree. These tortoises were allowed to reproduce freely up until 1976, so there is also a rabble of 36 offspring – most the result of undocumented pairings.

Between 1998 and 2000, the Yale geneticists sampled the DNA from these unidentified tortoises. Many had genetic sequences linking them to relatively robust tortoise populations on Santa Cruz and Isabela. Sadly for Lonesome George, none of them looks like it is a Pinta tortoise.

The use of genetic markers is becoming increasingly common in conservation. In the late 1990s, international concern over the expanding trade in wildlife parts led Canadian zoologists to purchase 21 penises from Chinese medicine shops in China and North America. These had supposedly come from legally hunted pinnipeds – seals, sea lions, fur seals and walruses. The researchers suspected that some of the penises were from illegally hunted species.

They plugged the genetic sequence from each penis into DNA databases. The results were both alarming and surprising: one of the penises seemed to be from the highly endangered Australian fur seal, a species that cannot be hunted legally. Another was from a bull (of the cow variety). Clearly, parts of all sorts of species (some of them endangered, some of them common) are finding their way into this lucrative market.

Edward Louis, of Omaha's Henry Doorly Zoo, who raised

questions over Lonesome George's identity, has performed this kind of forensic analysis on more than 400 captive tortoises, mostly in North America. Unfortunately, it's impossible to assign any of them to the Pinta population with any certainty, he says. Just as those studying shell shape only have the one-eyed female to go on, so Louis was looking for a genetic match with one individual – Lonesome George. Even so, says Louis, none of the animals had a genetic sequence that comes close to George's, so they are unlikely to have come from Pinta.

There are still many more captive tortoises to sample. Leading the way is Michael Russello, a postdoc in the Yale laboratory. There are several giant tortoises of unknown provenance in Quito Zoo and elsewhere on the Ecuadorian mainland, he says. If there really is a Pinta female surviving outside the Galápagos, this is where it's most likely to be found.

Russello is also in touch with Petr Velenský, the curator of lower vertebrates and invertebrates at Prague Zoo in the Czech Republic. There are two Galápagos giants in Velenský's custody and one of them – according to Peter Pritchard – looks suspiciously like a Pinta tortoise. Velenský is not so sure. The zoo records reveal that this individual arrived in Europe as a youngster in 1974, suggesting it probably hatched in the 1950s. It's therefore unlikely to have come from Pinta, which by that time had lost almost all its tortoises.

In any case, this Czech-based tortoise is yet another male – Tony. So even if his DNA does link him to Lonesome George, he's unlikely to be the saviour of the Pinta subspecies without some seriously futuristic interventions (more of which soon).

The Galápagos giant tortoise diaspora goes beyond zoo animals and museum pieces. There are creatures in private hands. Such animals are not always easy to gain access to.

In his 2002 book *Spix's Macaw*, Friends of the Earth

director Tony Juniper describes the challenges of getting private collectors to cooperate to save this stunning blue parrot from extinction. In late 2000, the last known wild Spix's macaw disappeared from a remote creek in the north of Brazil. The future of the species lay in getting captive animals together, not easy when a young bird might go for $30,000 on the black market.

There were then around 60 captive individuals – a few in São Paulo Zoo in Brazil, a couple at a reserve on the Canary Island of Tenerife and several in aviaries in Switzerland, the Philippines and Qatar in the Persian Gulf. Most were captive born and raised, had never seen life on the outside and many were closely related. The health of the captive population would be improved if private collectors could be persuaded to let go of their birds. Rumours circulated that there might be as many as 28, some allegedly held by the likes of Imelda Marcos, shoe fetishist and wife of the Philippines' dictator Ferdinand. 'Most of them had been bought privately – and mostly illegally – for large sums of money, and were thus jealously guarded prizes, the centrepieces of private collections', Juniper explains.

In 2003, one – a male called Presley kept as a household pet in Colorado – returned to Brazil to join other captive Spix's macaws. A press release from Birdlife International quipped: 'Presley can hopefully offer an invaluable boost to the species' survival and leave the very limited gene pool all shook up.' Whilst it's looking increasingly likely that some of these captive birds will one day be reintroduced into the wilds of Brazil, efforts to find out about and incorporate the rumoured birds into the breeding programme have been thwarted by the intensely private nature of these collections.

Private collectors of giant tortoises seem to be a lot more cooperative – for lots of reasons. First, most of the tortoises in private collections are not illegal. Few, if any, will have been smuggled from the Galápagos; most will have been born in captivity from tortoises collected at least half a century ago before such trade was outlawed. Second, for many collectors, a giant tortoise is a giant tortoise. If one of their animals did, in fact, turn out to be a Pinta female, they'd probably be more than happy to hand it over to the CDRS in exchange for a replacement.

There are several big private collections of giants in the US. One is at the Caloosahatchee Aviary and Botanical Garden in southwest Florida. Some of these tortoises are old, collected by the New York Aquarium's Charles Townsend in 1928. Russello has been allowed to sample these animals' DNA. Analysis will reveal whether any might help conservation efforts in the Galápagos and improve breeding at Caloosahatchee by showing which of their beasts are most suited to each other.

Of course, there is a slim chance that there are some private collections we don't know about. Like the search for the privately owned Spix's macaws, locating such collections will be tricky. Getting the owners to give up their precious pets could take years of bitter legal wrangling. Seeing them returned to the Galápagos would be nigh-on impossible. So it's time to return to the islands themselves.

Chapter 9
WILD AT HEART

The real reason for keeping a species turning over in captivity must be that one day it will be returned to the wild. The World Conservation Union calls this phase of a conservation initiative 'reintroduction', defining it as 'an attempt to establish a species in an area which was once part of its historical range, but from which it has been extirpated or become extinct'.

The most celebrated reintroduction is that of the Arabian oryx, a desert-dwelling gazelle from the Middle East, poached to extinction in the 1970s. Happily, a few animals in captivity had been brought together in Phoenix, Arizona in 1963. These mated, had young and 10 animals were reintroduced to Oman in 1982. This small herd was protected by members of the nomadic Harasis tribe and all appeared to be going well. In 1996, following further reintroductions, there were more than 400 individuals in the wild where once there had been none.

On paper, the oryx is a perfect species for a reintroduction attempt. It lives in a simple habitat, has very few natural predators, is tough and can take good care of itself; the main threat it faces is from hunting, which can be stopped (in theory). Despite ticking these boxes, the initial success of the Oman reintroduction did not last. As animal numbers rose, so did the incidence of poaching. In early 2003, the population had collapsed back to just 106 individuals, of which only six were female. The focus for oryx reintroduction shifted from Oman to Saudi Arabia, where there are two large herds.

Similar resilient attributes make giant tortoises suitable for reintroduction, yet they too present a slew of challenges. Not least, reintroduction assumes that there's something to reintroduce. In the early years, the focus was on transporting eggs from the wild, incubating and rearing the hatchlings in captivity and then returning them to their native island. This began in 1965 under Charles Darwin Research Station (CDRS) director Roger Perry.

Today the tortoise-rearing centre at the CDRS on Santa Cruz is a hive of activity. Godfrey Merlen summed up the bustle in his 1999 coffee-table tome *Restoring the Giant Tortoise Dynasty*:

> There are hundreds of tortoises and they come in all sizes. The smallest, a few centimetres long, are inside the pens, where they are dozing in the sun, clambering on a pile of small stones in the center, where their water supply is found. Outside are the big ones, some in excess of twenty-five centimetres and on the verge of being repatriated to their island of origin.

Tourists who gawp into these enclosures might be forgiven for thinking that rearing the reptiles is a cinch. Actually, this triumph took years of trial, error and careful experiment.

Initial efforts focused on Pinzón where black rats were eating up every young tortoise that managed to hatch; CDRS staff had not seen any on the island younger than about 30 or 40 years old. Something had to be done before it was too late.

Scientists decided to bring back eggs from Pinzón to incubate them and rear the hatchlings in the safe confines of the research station. This was harder than it sounds. Tortoise eggs are delicate things. CDRS biologists excavated nests on Pinzón, took out eggs and transferred them to metal cans. They sprinkled sawdust around the precious cargo to help it survive the journey along rough trails and several hours by boat back to Santa Cruz. There, they unloaded the eggs into custom-built incubation chambers, covered them with fine sand and under carefully controlled conditions let nature take its course.

Between three and eight months, the hatchlings begin to break free from their shells, a process that can take several days. Occasionally, a hatchling needed a helping human hand to get out. With their shells still soft, the baby giant tortoises spent several days in a dark box, living off the remains of the yolk sac to which they were still attached. Once their shells hardened and the yolk sac shrivelled up, they were moved to rearing pens fashioned from converted birdcages. After about four or five years steadily growing in captivity, the CDRS scientists reckoned that their youngsters were big enough to fend off the advances of the rats on Pinzón.

By the late 1960s, hundreds of tortoises were being reared at the research station. On 11 December 1970, the first batch of 20 captive-reared animals was returned to Pinzón. They took to their new surroundings without any fuss, feeding happily within minutes of being set down. There was no sign of rat attacks, and they put on weight rapidly. It was a great success. Further investment in facilities and the extension of the programme to include other endangered species seemed warranted.

Next in line was the Española tortoise. In August 1963, when combing Española for signs of tortoise life, three research station staff came across a single animal. It was clearly having a hard time making ends meet. 'When the tortoise was found, it was feeding on a fallen *Opuntia* [cactus] in company, and in competition, with 15 goats', wrote David Snow, then director of the CDRS.

Thereafter, expeditions to the interior of Española came across more tortoises. As these were discovered, they were shipped to the CDRS on Santa Cruz. The last tortoise to be found on Española – a female – turned up in 1972, the same year that Lonesome George was brought into captivity. 'The posterior slope of its carapace is covered with lichens, an indication that it has not reproduced for decades', someone noted at the time. With the addition of this crusty animal, the Española captives numbered just 14 animals – 2 males and 12 females.

This presented a new challenge for the CDRS. The Pinzón programme had brought eggs not tortoises from the island. There were no eggs on Española to bring. It was clear that the last few Española giants were struggling to find each other on their island and breeding in the wild had ceased altogether. The future of this subspecies depended on getting them to reproduce in captivity.

In the first season, things did not go well. Once a female tortoise has laid her eggs, she gently rearranges them, nudging them into a single layer with her rear feet. She then covers the eggs with soil. During these two stages – rearranging and covering – a lot of eggs were lost.

Some felt that they were being damaged as they dropped into the nest cavity. CDRS scientist Craig MacFarland, who became director in 1974, thought otherwise. Reptile eggs are encased in a thick, gelatinous fluid; they ooze rather than drop, cushioned from any impact. The real problem, he argued, was that the ground in the research station was too stony for nest-building. 'The soil should be relatively fine and form an adhesive but workable mud when wetted by the copious urination of the female during the excavation', he and his colleagues wrote in *Biological Conservation* in 1974. Far fewer eggs got damaged when the soil was changed from coarse to fine.

Baby tortoises were still a long way off. Several females dug down into nest chambers that already contained eggs, destroying whole clutches laid the night before. To avoid this, MacFarland and his CDRS colleagues dug up fresh-laid eggs to clear the space for another female to lay. Dusting away the soil like archaeologists, they carried the delicate day-old eggs some 200 m to the incubators, placing them alongside the eggs from Pinzón.

Tortoises can lay two fertile clutches each year. For most of the Española females that tried this double-whammy, things didn't work. Nevertheless, by 1975, the breeding programme had produced over 80 baby Española tortoises. Seventeen – born in 1971 – were even old enough to be returned to the wild.

Figure 9.1 Captive-bred tortoises are reintroduced to Española

When animals are ready for reintroduction, the question remains: is the ecosystem ready to receive them? For Galápagos giant tortoises, this generally means: is the island still overrun with goats and pigs? If so, these pests must be controlled or ideally eradicated.

Ridding a place of such creatures takes time, effort and money. The Galápagos National Park Service (GNPS) spent more than 30 years waging a war against Pinta's rampant goats. Shooting began in 1971, the same year that Vagvolgyi spotted Lonesome George. There were so many goats on the island – an estimated 20,000 – that at first the work was easy. Undaunted, the animals bred almost as fast as park rangers could remove them. In the 1970s alone, wardens shot an estimated 41,390.

Sadly, their effort was not constant. The number of trips each year varied, as did the manpower and length of stay. The hunters were armed with 0.22 calibre rifles and sometimes had untrained dogs, which probably didn't help much. There was no counting system. Today wardens on foot collect the tail of each goat they shoot and aerial hunters take a GPS reading for every animal that goes down. Back then there was no such record. So we can't be sure of the exact number of goats killed.

In an effort to slow the birth rate, the hunters targeted females. Plus, males tend to herd together in the absence of females, and these groups of bachelor goats could then be picked off with ease. Whatever the exact number, the rotting goat corpses strewn all over the island suggest that the massacre was massive. In 1985, there was a huge push, but only eight animals were shot. These, it was believed, were the last ones. 'A meticulous search, aided by dogs, allowed us to conclude that all surviving goats had been killed', boasted that year's annual report of the CDRS.

Meanwhile, Española was being prepared for the return of its tortoises. Goats hadn't taken over as badly as on Pinta. Still, the park service had to hunt down and shoot nearly 3500 animals. When the last was dispatched in 1978, the repatriated tortoises had free rein of the island once more.

Back on Santa Cruz, the breeding from the 14 Española tortoises was going from strength to strength. It was a real bonus that so many were female: more females means more eggs. It was of some concern that only two of the animals were male. Then in 1977, another male, 'No. 21', turned up in an unlikely place.

At that time, San Diego Zoological Park had an impressive collection of about 25 Galápagos tortoises. Thomas Fritts, who had just started scouring zoos for a mate for Lonesome George, noticed that one of the San Diego males was different from the rest. He had a definite saddleback while the others had domed shells, he kept himself apart and, when he did interact with others, he could be pretty aggressive. Everything suggested that this tortoise was from Española, Fritts decided.

The zoo's records backed up his suspicions: there were originally seven Española tortoises at San Diego, six of which had died. The shells of the unfortunate exiles had been rolled down a ravine at the back of the zoo. The antisocial saddleback – No. 21 – had to be the last remaining Española tortoise in the collection. MacFarland, by now director of the CDRS, agreed.

The zoo authorities accepted the evidence and let No. 21 return to the Galápagos to join his fellow Españolas. The first leg of his journey was by aeroplane to Guayaquil. From there, he was to be shipped to the Galápagos on board a tourist vessel. Fritts was anxious about this option, as it would lengthen the trip by several days. His fears dispelled when he recalled that 'during the last century thousands of tortoises (perhaps including the ancestors of No. 21) had lived for months in the holds of whaling vessels during their tragic journeys from the islands'.

No. 21 docked at the CDRS on 8 August 1977 and lumbered out of his container into the bright sunshine beating onto the jetty. Within days, MacFarland was handfeeding him the fleshy leaves of the Galápagos prickly pear. Soon, he was avoiding visitors to the isolation corral in which he was temporarily housed. Later that year, No. 21 completed his extraordinary journey when he was freed from his quarantine and allowed to join the other Española tortoises.

The repatriation of No. 21 was a spectacular success. He quickly assumed the role of dominant male and injected some much-needed genetic variation into the next generation. No. 21 was such a randy beast that his custodians dubbed him 'Macho'.

Now that Española was goat-free, the reintroduced tortoises seemed to be thriving. In November 1990, there were indisputable signs that animals on the island were mating once more. Wardens reported seeing one female completing her nest and another with a muddy rear that had probably just finished digging out hers. Since all reintroduced animals are given individual markings, it was possible to work out that these females hatched on Santa Cruz 17 years earlier in 1973 and were set free on Española in 1978.

When the wardens moved in to get a closer look, they found four nests and signs of other nesting attempts. They also came across the remains of two one-month-old hatchlings that had been devoured by hawks. In spite of the predation, the conservationists were upbeat: 'Their appearance on Española is a major indicator of the ultimate success of this long-term programme and the possibility exists that other young have escaped the notice of hawks and scientists alike', they wrote in *Noticias de Galápagos*. The tortoises were settling in to their new surroundings pretty quickly.

This isn't always the case for a reintroduced animal. Some – particularly social creatures – need considerable preparation before they can be released into the wild. Returning golden lion tamarins to the Brazilian rainforest has been tough. In 1983, 15 captive-born monkeys were sent to South America. Four died in quarantine before they were released. In the middle of 1984, the remaining 11 animals were transferred to a reserve and held in large cages to acclimatize. These animals were released into the forest but within a year only three remained, the rest either dead or rescued by researchers after getting into difficulties.

The captive-bred tamarins weren't fit for life on the outside – they didn't like trees and preferred to wander along the ground. A snake snatched one and a dog savaged another. What was missing from this early effort was training. The tamarins needed to be taught how to climb, how to find food, where to sleep and how to interact. The solution was to bring an experienced wild tamarin into the group to teach the captive-born newcomers.

Thankfully, the Española tortoises didn't seem to need this kind of schooling. Their captive lifestyle was not so different from life in the wild: the equatorial climate, the rocky terrain and vegetation in their pens set them up well for returning to Española. When the 1000th was repatriated in 2000, there was a big celebration.

A couple of years later, molecular biologists sounded a note of caution. Genetic analysis of the repatriated animals revealed an unforeseen problem: the 15 founding tortoises were not contributing equally to the reproductive effort. Michel Milinkovitch, a geneticist at the Université Libre de Bruxelles in Belgium, working with the Yale team, took blood from over 100 animals on Española to probe their parentage. The results were surprising.

Macho, the San Diego male, is the father of around 60% of the baby tortoises leaving the CDRS; the other two males are clearly not pulling their reproductive weight. Impressive as it is, Macho's zealous contribution to the captive-breeding initiative could actually have undermined it. Chances are that many of the natural pairings now taking place on Española are between his sons and daughters, between half-brothers and half-sisters.

As we've already seen, inbreeding is something that conservationists are usually wary of. Most animals inherit two copies of each gene – one from each parent. If they happen to get a faulty copy from one parent, it's not too serious because the other parent has it covered. If, however, the mother and father are close relatives – say half-siblings – they will be passing similar genes to their offspring. So if the father serves up a dud copy, there's a greater chance the mother will too.

As was probably the case for the dwindling population of Florida panthers, this can result in offspring with health defects and poor fertility. It also reduces the genetic variation in the population. This makes the group more vulnerable to change – an infection or some weather that might normally kill a few animals could wipe out an inbred colony.

Richard Frankham, a conservation geneticist at Macquarie University in New South Wales, and his colleagues have used fruit flies to explore the effect of inbreeding on the ability to fight off disease. They had two groups of flies – one inbred, the other not – and exposed each to a toxin and an infection. The inbred flies were hardest hit. 'Wildlife managers should strive to minimize inbreeding and loss of genetic diversity within threatened populations and to minimize exposure of inbred populations to disease', concluded Frankham's team.

Figure 9.2 The northern elephant seal made it through a tight population bottleneck

That said, in some species a bit of inbreeding seems to pass without incident. Northern elephant seals squeezed through a population bottleneck after relentless hunting during the 19th century reduced the worldwide population to just 20-odd animals on Guadalupe in the Pacific. Early molecular work suggested that they were all very close relatives. Now there may be as many as 150,000 descendents of this tiny founding group.

The survival of the Mauritius kestrel is even more startling. Today, there are nearly 1000 descended from just two pairs, the nexus of a careful breeding programme. Habitat destruction and widespread use of fertilizers like DDT had caused the population to crash. Some suggested that the species should be left to go extinct. In 1979, Welshman Carl Jones waded in, determined to revive the endangered kestrel.

He took eggs from the two wild nests, stimulating both pairs to lay another clutch. Jones incubated the first clutch in his laboratory. Some of these captive-born chicks he inserted back into nests to be fostered by the wild kestrels. Others he put in artificial nests and fed until they could forage for themselves. The last kestrel was freed in 1994, since when the population has bounced back. Most of the birds have the same genes. But so far, this hasn't stopped them.

A single female may even be enough for a population to survive. During the cold winter of 1948, an ice bridge formed across Lake Superior, connecting one of its islands – Isle Royale – to the Canadian mainland some 24 km away. It's thought this is when a small pack of wolves made it to the island. Genetic fingerprinting suggests that the 30 or so now on the island are all descended from just one female. These Isle Royale gray wolves do show all the downsides of inbreeding, like small litters of which many die, but so far, at least, they have escaped extinction.

Perhaps then Macho's virility will not be too much of a problem for the future of the Española giants. Linda Cayot, who managed the Española programme for much of the

1990s, isn't worried, pointing out that inbreeding is par for the course for giant tortoises. As she puts it: 'It's more than likely that all the tortoise populations on Galápagos started with just a few founding individuals.'

A bottleneck, inevitably followed by a bit of incestuous mating, may result in damaging genetic combinations, ill health and frequently death. If, however, a population avoids extinction, it may find itself purged of some troublesome genes. Life on the other side could be rather sweet.

Back on Pinta there was bad news. The 1985 report that goats had been eradicated turned out to be premature. Just three years later, five goats were found and six in the year after that. Then in July 1990, after a visiting botanist had spent 30 days on Pinta without seeing a single sign of a goat, the island was declared goat-free once more: 'So it seems that on the last trip by GNPS and CDRS personnel in February this year, they killed the last animals', was the bold conclusion of a research station report. Yet again this assessment was too hasty. Five years later, park wardens found and shot another 25 animals. Like Australia's rabbits, cane toads and fire ants, the Pinta goats were proving almost impossible to stamp out.

Enter the 'Judas goat'. Capitalizing on the animals' herding instinct, Galápagos researchers fitted a few with radio collars and released them. The idea is this: collared animals head off into the inhospitable and inaccessible parts of the island and hook up with other goats. You return a few months later with a radio tracker, locate the Judas goats and shoot all the animals with them. You do not shoot the Judas goat. You go away, come back a couple of months later and do the same. You keep doing this until you radio track the Judas goat and find it alone. Then you shoot the Judas goat.

It must be pretty disconcerting being a Judas goat. Just

when you think you've made a new friend, bang! It slumps to the ground. During 1999 and 2000, 28 roamed Pinta's mountainside, experiencing the worst kind of Groundhog Day as new acquaintances were dispatched before their eyes time after time. By 2003, these bewildered animals appeared to be the only goats left on Pinta.

The snag was how to be certain. The park service was keen to avoid a repeat of the hasty 1985 and 1990 claims. They planned a thorough search of the island to say, once and for all, whether the Judas goats had been successful. This was when Peter Pritchard approached them to see if he could search for tortoises.

During the joint GNPS–Pritchard expedition in 2003, the wardens did find one goat. However, there's good reason to believe that it had only recently arrived. Unlike the black and white goats that have clung to Pinta's craggy slopes for years, it was brown. There were also rumours that disgruntled fishermen were trying to sabotage the eradication effort. 'From what we can gather it's malice', says CDRS scientist Karl Campbell. They shot the goat.

This would not be the first time that local people have attempted to interfere with conservation work. In the Galápagos, there is a constant threat of such sabotage. The island of Marchena, for example, had been free of goats for over 30 years. But since 2000, they've returned. Which just goes to show that eradication is not the end of the fight against an introduced species. Regular monitoring can act as a stop-gap to ensure that the problem does not recur. The long-term solution is to foster a cooperative spirit between the different interests in the area.

The eradication of goats from Pinta is a great achievement, but it was just part of a much greater exercise – the Isabela Project. In 1997, CDRS scientists devised an ambitious plan to try to remove an estimated 100,000 goats from the vast expanse of northern Isabela.

Ready for a new challenge, Linda Cayot headed up the new initiative. She built the Pinta eradication into the pilot phase of the project, giving wardens the task of seeking and destroying a very low-density population. They honed their skills and they seem to have succeeded, clearing the way to tackle Isabela. While the scientists deployed fewer than 30 Judas goats on Pinta, there are already over 500 Judas animals on Isabela snitching on their unsuspecting chums. The tremendous success of this initiative has given reason to expand goat killing to include the southern reaches of the island.

So Pinta is now goat-free. It's ready to receive tortoises. While Lonesome George refuses to breed, they will have to be from a different island.

There is a precedent. Giant tortoises were an important feature of most of the islands in the Seychelles and Mascarenes in the Indian Ocean. As soon as settlers killed off these dominant herbivores, the plant life ran rampant, particularly near the coast. What's more, the paths made by the tortoises channel rainwater down the mountainside. No tortoises meant no channels. The islands were changing. So conservationists brought giant tortoises from one island, Aldabra, to repopulate others like Mauritius and Curieuse. The effect was dramatic, quickly restoring some sort of balance to the off-kilter ecosystems.

Some think this idea could be taken further. One of them is Josh Donlan, a conservation biologist at Cornell University in New York State, who has helped with the goat eradication on Pinta and Isabela. In 2005, he and several colleagues published a commentary in *Nature*, suggesting that it's time to return North America to something like it was before humans came on the scene around 13,000 years ago.

Back then, the grasslands teemed with deer-like pronghorn, medium-sized tortoises, wild horses, camels and five species of elephant-type creatures, while roving predators like American cheetahs and lions kept the whole lot in check. Most of these species are long extinct, but Donlan reckons modern equivalents could be wheeled in to replace them. His radical vision is to use North America as a playground for the endangered. This would restore complexity to a landscape currently dominated by pests and weeds; African and Asian elephants, for example, could help knock down woody plants threatening the western grasslands. It would bring economic benefits. It would also improve the long-term prospects of many of these endangered creatures; in Donlan's world, if the lion were to go extinct in Africa, the North American population could act as back-up.

Pinta's plant life has bounced back now that the goats have gone. Indeed, without any major herbivore, it's become seriously overgrown. Many feel strongly that Pinta needs one to keep its ecosystem on track. Returning a subspecies of tortoise should be tried, says botanist Ole Hamann. They would open up the dense vegetation, helping light-loving herbs and grasses to thrive and maintaining the diversity of the island's plants, he says. It's also a reversible experiment. 'If, in 20 years time, somebody decided that this wasn't such a good idea, you could probably get the tortoises off quite easily.'

Goat exterminator Campbell has put together a proposal to the GNPS to return giant tortoises to the island. Because the Española tortoises appear to be most closely related to Lonesome George, they will likely be the stand-ins of choice.

It has been suggested that Lonesome George should join them. For the moment though, the research station on Santa Cruz is the best place for him. While he remains in captivity, there is always a chance (albeit remote) that the scientists around him will fathom out a way to reproduce him.

Chapter 10
FAKING ORGANISMS

If geneticists do manage to identify a Pinta female for Lonesome George, it would be a triumph. If they don't, it would still be worth sticking one or two of the more closely related Española females into his enclosure. There's no guarantee (even with a bit of manual encouragement from someone like Grigioni) that George would show any more interest in them than he has in the Isabela females currently in his pen. If he really turns out to be as disinterested in sex as he appears, what then?

It might be worth thinking about artificial insemination, particularly useful when animals are refusing to mate. Quite simply, a human acts as go-between, ferrying sperm from male to female. This middleman (or woman) can sometimes offer the additional service of placing the sperm in the best place and at the best time to meet an egg. If fertilization does occur, the go-between can hang up his or her lab coat and the female does the rest.

Artificial insemination has the advantage of being pretty low-tech. Unfortunately, it still relies on natural fertilization, which is not a given. The female's reproductive tract is a hostile place and many sperm are needed for there to be a good chance that a single egg will be fertilized.

The plan might be to inseminate an Española female. The snag, of course, is that this is all rather hypothetical unless George can be coaxed to yield some sperm. If a genital massage à la Grigioni doesn't do the trick, it might be time to turn to electroejaculation.

—

Electroejaculation is often used on captive animals because it's relatively simple and predictable. It is particularly helpful where the creature in question is so dangerous that nobody in their right mind is going to seriously consider taking a male aside and engaging in a bit of friendly manual stimulation.

In his 2002 report to the CDRS, Thomas Fritts pushed the idea of electroejaculating George: 'Given the minimal risk that such an effort would pose and the high visibility that it would receive in public and conservation communities, I can envision that visiting experts could be enlisted to perform this work at no or minimal cost', he wrote.

What would be involved? Electroejaculation usually requires the subject to be anaesthetized (although, as it turns out, not reptiles). The electrified probe stimulates the testes (thus releasing sperm) and so-called accessory glands (thereby ejecting other important secretions). An ejaculate, of sorts, emerges.

Obviously, the result is not what the animal would have produced naturally. We should perhaps call it an 'electro-ejaculate', an emission which tends to be relatively flush with accessory fluids and contains poorer quality sperm than a natural ejaculate. Nevertheless, for reproducing some exotic species it's the only option.

Electroejaculation has been used to collect semen from pandas, tigers, koalas and rhinos. It is also used on wild elephants, where Hildebrandt's massage is simply not an option. The animal is immobilized with a tranquillizer dart, out comes a customized (large) rectal probe, ultrasound is used to guide it to the right place and the voltage is steadily increased.

This practice, as ever, is best developed for mammals. For reptiles, things are relatively hit and miss. Back in 1980, Carrol Platz Jr, then at Texas A&M University, helped devise a protocol for reptiles, adapted from his experience with mammals, where the voltage is increased until it causes extension of the animal's hind legs. The electrical stimulus needed to reach this point is known as 'peak voltage'.

First, a tortoise's penis must be exposed. As we've seen, this is done by gently massaging its tail. Alternatively, the animal is held off the ground, until eventually the muscle that holds the penis inside the tail relaxes and the penis drops out. The probe is then inserted into the tortoise and the electrical current applied. For reptiles, peak voltage is assumed to be the point at which the penis becomes rigid.

All reasonable efforts must be made to avoid simultaneous ejaculation and defecation, as any mixing can wreck a sperm sample. In mammals, this is not much of a problem, because with judicious penis direction, these emissions can be kept apart. Reptilian anatomy – the multipurpose cloaca – complicates things considerably. 'That was probably the biggest problem we had', says Platz Jr. His team discovered that a good sample can be come by if they flush out the reptilian digestive tract with saline first.

This method has been used to collect sperm from several reptiles. Honolulu Zoo in Hawaii has been home to a captive population of endangered Madagascar ploughshare tortoises since the early 1970s. For most of their first decade in residence, there was no sign of breeding. With the arrival of a female from San Antonio Zoo in Texas, things suddenly started to happen. In just three years, the new recruit had laid six clutches and a total of 25 eggs.

Alas, none of them hatched. Perhaps the female was not getting enough sperm from the males to fertilize her eggs, reasoned the zoo biologists. They began to think about electroejaculation. If only they could get hold of some sperm, they could do the rest through artificial insemination.

Having honed their electroejaculation skills on a few workaday tortoises, they managed to collect a small amount of apparently good-quality sperm from a male ploughshare.

Twelve days after inserting it into the San Antonio female, she laid a clutch of seven eggs. None of these eggs hatched, but one contained an embryonic tortoise. This was the first sign of fertility. A few months later, the female successfully hatched her first baby since arriving in Hawaii.

Platz Jr and his co-workers also experimented with Galápagos giant tortoises in captivity at a couple of zoos in Texas. They got sperm. 'There were good concentrations and they were very motile', Platz Jr says.

All the indications are that electroejaculation could work for Lonesome George. Still, any medical intervention runs a risk to the patient – inserting a probe into an animal's rectum and switching on an electrical current probably more than most. When the patient is one of a kind like Lonesome George, it would be a brave call indeed to give the go-ahead. Linda Cayot says: 'Electroejaculation is not something they would ever do.'

If George could be coaxed to ejaculate (naturally or by shock tactics), it would make sense to divide up the sample and freeze it in several batches. These 'cryopreserved' shots could be defrosted as and when needed for a series of artificial insemination attempts.

Putting sperm on ice is tricky. Freezing and unfreezing are brutal processes that have a habit of destroying the integrity of cells, explains reproductive biologist Bill Holt of the Institute of Zoology in London. There are only a few cases where frozen sperm cells from endangered species have gone on to produce live offspring. These include the cheetah, the Mohor gazelle, Bactrian camel and a handful of other mammals.

A memorable success is the black-footed ferret. This cute weasely creature, native to the prairies of North America, was thought to have become extinct because of the fragmentation of its habitat and the eradication of its main source of prey, the prairie dog. In 1981, a ranch dog killed a lone black-footed ferret in Wyoming, which triggered a search for a family of ferrets. A small population of what was then the world's most endangered carnivore was found.

After a series of disease outbreaks, the ferrets were in such peril that between 1985 and 1987, the last 18 animals were trapped and brought into captivity. From these individuals, with the help of cryopreserved sperm and assisted reproduction, the National Black-Footed Ferret Conservation Center has produced more than 4000 animals and gained an extensive knowledge of ferret biology.

One early innovation was the 'black-footed ferret genome resource bank'. This is a repository of frozen sperm from the most genetically valuable males, especially those that were failing to reproduce by natural breeding. With the expanding population of ferrets housed at multiple institutions to protect against a disease epidemic, it made sense to move sperm rather than males from facility to facility to maintain maximum genetic diversity. There are now several hundred black-footed ferrets back in the wild.

This exception apart, freezing is bad for the viability of sperm, eggs and embryos. Cryobiology is so fraught with difficulties that conservation biologists like elephant supremo Hildebrandt are forced to extraordinary lengths to carry out a fresh insemination. Attempts at freezing elephant semen have been disappointing, which means that once Hildebrandt has an ejaculate in the bag, there are but hours to get it to the female. With immaculate planning, he has successfully couriered elephant semen from a male in London to a female in Toronto – some 5500 km. Not even an African bull elephant with its legendary penis is capable of holding down such a long-distance relationship without help.

Figure 10.1 Sperm cryopreservation helped save the black-footed ferret

Few people have even experimented with the conditions needed to freeze reptilian semen. Those that have had little success. 'The best I ever got was 10% recovery in an Angolan python', says Platz Jr. Struggling with the whims of reptilian semen, he wandered away from academia to set up a niche business that collects, freezes and ships ejaculates from prize-winning dogs to breeders wanting to add another trophy to their cabinet. There is a role for cryobiology in conservation, concludes Holt, but it is limited. It should only be a supporting technology for other more direct conservation measures, he feels.

For the moment then, sperm from Lonesome George would have to be put straight into a waiting female. Interestingly, she may be able to pull off what the biologists can't, acting as a 'natural freezer'. Females of most egg-laying species can store sperm and keep them viable, probably for many months. Most mammals (including humans) aren't able to do this and sperm die within days of insemination. The storage feat is achieved by specialized folds in the female reptile's tract, which protect and nurture sperm until they are needed. These microscopic structures have been best studied in birds. The turkey, for example, holds the avian record for storing sperm: one bird laid a fertile egg more than 100 days after sex.

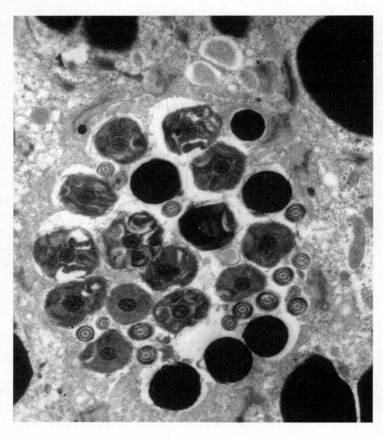

Figure 10.2 A slice through several sperm in the storage gland of a female box turtle *Terrapene carolina*

What little research has been done suggests that reptiles are ace sperm storers too, possibly even managing to keep it in working order for several years. A preliminary study on the eastern box turtle *Terrapene carolina* found females could keep sperm viable for at least four years. A single artificial insemination using Lonesome George's sperm could therefore result in offspring several years down the line.

Of course, this means that if the female mated with any other male in the interim, a paternity test would be needed to confirm George was the father. Indeed, DNA profiling of desert tortoise *Gopherus agassizi* hatchlings shows that many females have clutches of multiple paternity. In one female, nearly half the baby tortoises were fathered by a male that she had mated with over two years before she laid her eggs.

The success of artificial insemination relies, to a large degree, upon having lots of good-quality sperm. What if George isn't producing many or even any active sperm? The game isn't up.

One possibility is that there's something amiss with his diet. Soon after George's arrival at the CDRS in 1972, he began to put on weight. Life in captivity was so much easier than life on Pinta. His new sedentary existence, combined with the lavish attentions of a doting warden, drew him dangerously close to corpulence. George was nearly falling out of his shell.

Being fat dents fertility. A study of over 1500 Danish volunteers found that overweight men had significantly lower sperm counts than those of normal weight. The reason is complex, but it almost certainly has something to do with hormones, which start to go out of balance when you pile on the pounds.

This may have been the case for George. In the late 1980s and early 1990s, Olav Oftedal, head of nutrition at the National Zoo in Washington DC, was in the Galápagos advising on the diet of captive land iguanas. While there, he took a look at the tortoises. Lonesome George had a swelling to his neck, Oftedal noted in a 1994 report to the research station. This suggested an enlarged thyroid gland, which might explain his weight problem and lack of sex drive. George might also be suffering from a deficiency of trace elements needed for sperm production. Oftedal recommended that he be put on a mineral-enhanced diet. The celebrity soon lost weight. As far as anyone can tell though, being svelte did not change his attitude to sex.

If George doesn't have many sperm to play with, there are still two more high-tech options available – *in vitro* fertilization (IVF) or intra-cytoplasmic sperm injection (ICSI). Regrettably for George, these are techniques best suited to mammals. Pulling off IVF or ICSI for a reptile wouldn't be far off a miracle.

The advantage of IVF and ICSI is that they work with very few sperm. The huge disadvantage is both require the collection of unfertilized eggs as well as sperm, which are then mixed in the laboratory. For IVF, these cells are crammed together in a dish, dramatically increasing the chance that a sperm will bump into an egg. If it doesn't, there's still ICSI. Here, a single sperm is injected into a single egg beneath a microscope, particularly useful if the sperm are having trouble swimming. This virtually guarantees fertilization.

IVF and ICSI are possible for humans because we've found out how to collect eggs: we know how hormones fluctuate over time, blood tests can reveal the levels of these hormones at a particular moment, ultrasound can monitor the growth of eggs and drugs can induce ovulation. Sadly, what applies to the females of one species rarely applies to the females of another. As David Wildt, head of reproductive sciences at the National Zoo in Washington DC, quips, 'a cheetah is not a cow'.

So all the sperm in the world, if Lonesome George could only provide it, would not improve the chances of successful

IVF or ICSI. An even greater challenge is working out how to collect unfertilized eggs from a female tortoise.

A good first step is to find out whether Galápagos tortoises are spontaneous or induced ovulators. In species that have spontaneous ovulation, females release eggs throughout their reproductive lives, even if there are no males around. Humans are spontaneous ovulators, and women go through a (relatively) predictable cycle that sees the growth of eggs in the ovaries, preparation of the uterus lining for the possibility of pregnancy, release of an egg or occasionally eggs (ovulation) and, in the absence of fertilization, the shrinking and shedding of the uterus lining (menstruation). This set-up works pretty well in a predictable environment, where males (and hence sperm) are relatively easy to come by.

In species exposed to wild seasonal fluctuations and where males are rarely encountered, it doesn't pay to invest in non-stop fertility. Here ovulation has to be induced by a particular set of stimuli. Tortoises are induced ovulators.

For most induced ovulators, eggs develop in response to environmental cues like food abundance, rainfall or temperature, but they are not released from the ovaries unless certain additional conditions are met. As a rule – and it's not a bad one really because it helps coordinate the rendezvous between sperm and eggs – copulation is often one of these conditions. This, in theory, should make egg collection easy: the cervix is stimulated and eggs follow. In rabbits, for example, just 200 milliseconds of copulation trips ovulation. Such a sure-fire response is unusual; more prolonged stimulation and other cues are often required. In mink, courtship alone can bring on ovulation; for mice, the trigger is the presence of the peculiar whiff of sexually charged male pheromones. In tortoises, we simply don't know – perhaps all these factors and more must conspire for eggs to be released.

Hormone analysis can help work out when eggs are about to be released. Admittedly, this is most often used to determine the best moment for artificial insemination but it can also help in the collection of eggs to be used for either IVF or ICSI. Scientists reserve a special place for luteinizing hormone (LH). This protein is produced in the brain and secreted into the bloodstream where it acts as a signal to the ovaries and several other organs. As a female's reproductive condition changes, so too do her levels of LH. In most mammals, an enormous peak in LH concentration triggers the release of an egg or eggs from the ovaries. This LH peak is an unmistakable sign that the time is right for insemination or egg collection.

Simple in theory. In practice, it means collecting daily blood samples from the female singled out for treatment, and these samples must be analysed immediately. It would be no good stacking up a batch of blood samples to inspect them all in one go: the LH peak must be caught and acted on swiftly. Timing is crucial.

Once again elephants offer a perfect illustration of the peculiarities of each species. In 1995, it was discovered that in a single reproductive cycle, female elephants produce two LH peaks rather than the one common to most other mammals. These two spikes are separated by 20 days (give or take a day), the second apparently stimulating ovulation. Suddenly, this unique double peak explained why some inseminations (those that coincided, just by chance, with the first LH peak) produced no results: no eggs were being released so no fertilization could occur. The function of the first LH peak remains a mystery, but now that reproductive biologists are aware of this unusual hormonal pattern, they can exploit it. Once they spot a first peak, they know that in 20 days, they need to be back with some sperm. If the LH calculation is accurate, the insemination goes to plan and the sperm quality is good, fertilization is almost guaranteed.

It could be useful to measure LH in Lonesome George's putative females. The hitch is that the reptilian version of the hormone has a different chemical structure to the mammalian version. So the method used to measure LH levels in elephants and other mammals doesn't work for reptiles. There are other hormones that we can detect, and thanks to a doctoral thesis in the late 1990s by Beatrix Schramm, then at the University of Zurich in Switzerland, we do know something about the reproductive patterns of female Galápagos tortoises. Should anyone ever get round to trying artificial insemination in these animals, her research will be invaluable.

Working at the CDRS, Schramm measured how the concentration of certain steroids varies throughout the year. To immobilize the tortoises, Schramm tipped them over and rested them in a car tyre. She took blood samples, which were flown out to San Diego for detailed analysis, and used ultrasound to clarify when egg follicles began to develop in the ovaries. This work has laid some groundwork for anyone trying to get these follicles out of a female and into the laboratory for IVF or ICSI. It doesn't address the fact that you'd probably have to subject the female to an operation to get at them. 'I would not recommend it at all', says Schramm.

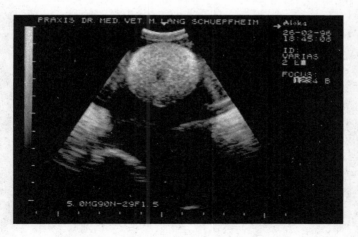

Figure 10.3 Ultrasound image of a Galápagos tortoise egg just before ovulation

As if all these hurdles weren't enough, there is one much more significant drawback of bringing fertilization into the laboratory. In most mammals, a lab-fertilized embryo is inserted into the uterus of a waiting surrogate female, and with some judgement and a good dose of luck a placenta will form. Egg-laying animals like birds or reptiles do not have placentas. Instead, a fertilized egg is packaged up with a yolk inside a shell and becomes relatively independent from its mother. The reproductive biologist forced down the IVF or ICSI routes with a reptile must therefore address what might reasonably be called 'the shell problem'. They must find a way of supplying the embryo with a yolk (or equivalent source of food) and giving it a shell (or equivalent protection).

With our current meagre understanding of reptilian reproduction, these are formidable challenges that will require some seriously futuristic interventions. For now, there's only one place for these and that's the final chapter.

Chapter 11
CLONES AND CHIMERAS

'Cloning could be considered when all other options have been exhausted', announce the panels around George's enclosure. They did it for Dolly; why not just create an exact replica of George, gene for gene? In theory, it's possible. It even sounds fairly straightforward. It isn't.

Cloning gets a lot of bad press. But in the natural world it has been around for billions of years. In fact, for most organisms it is the preferred mode of reproduction; why invest in sperm and eggs and waste all that time trying to find a suitable partner of the opposite sex when you could just pop out a copy of yourself? Simple organisms like bacteria, yeast and protozoa do it. More complex organisms, such as some corals and sponges, made up of millions of cells rather than just one, also go in for cloning from time to time. Many plants are professional cloners. There are even mammals like the nine-banded armadillo (which gives birth to identical quadruplets) and humans that are capable of manufacturing occasional clones (we prefer to call them twins).

That said, if the talk turns to cloning, chances are that it will be Dolly the Sheep that springs to mind, not bacteria, yeast or twins. Dolly may be the most famous clone born with assistance from humans; she is by no means the first. This title probably goes to a 19th-century sea urchin that spent its short life at the International Zoological Station in Naples, Italy. In 1894, German biologist Hans Driesch dropped a two-celled embryonic sea urchin into a beaker of water and shook it until the cells parted. The result was two, independent and identical sea urchins instead of one.

Another important early clone was a salamander hatched in a laboratory in early 20th-century Germany. The anonymous amphibian came about through an ingenious experiment by Hans Spemann, a young and gifted scientist who wanted to work out how a complex animal – with structures like skin, bones, muscles and eyes – can develop from a single cell. He reasoned that once sperm has met egg, the resulting cell (technically known as a zygote) must contain all the information necessary to build all the different structures in the adult animal. In 1902, while at the University of Würzburg, Spemann determined to find out whether this impressive property of a single cell is lost as it copies itself by dividing in two.

The animal he chose to study was a salamander because its eggs, rather like frogspawn, are large and easy to work on. He took a salamander egg and a salamander sperm and let them fuse to create a zygote. He put this single salamander cell onto a slide and moved it beneath the lens of his microscope. Peering into the eyepiece, Spemann made out the cell, a circular membrane containing a single nucleus that, he reckoned, must contain the blueprint to construct an entire salamander. He looped a strand of hair from his baby son around the cell like a noose. Slowly, he tightened. The fine hair was a good way to split the dividing embryo in two, and Spemann found that each cell could still make an entire salamander – one a clone of the other. As the nucleus divides, he surmised, the information it contains is preserved. This property turns out to be a feature of all cells taken from a very early embryo. Such embryonic cells are described as totipotent – literally 'all-powerful'.

At some point during development, cells lose this totipotency and become devoted to performing one task for the rest of their lives. This specialization is called differentiation. It's as if there comes a time when these young and frivolous totipotent cells must assume some responsibility, decide on their future and stick to it. Just as a child has the potential to become a train driver, a banker or a brain surgeon, so an embryonic cell could differentiate to live out its days as a muscle, a spleen or a nerve cell. And just as it becomes increasingly difficult to change career with age, so differentiated cells find themselves trapped, with the rest of their lives mapped out before them.

With some effort, it is possible to quit the job in the bank, go back to those heady days of youth and retrain. And in the second half of the 20th century, evidence began to accumulate that differentiated cells might be able to backtrack in a similar way.

In the 1950s, two developmental biologists in Philadelphia – Robert Briggs and Thomas King – devised a technique called nuclear transfer. Spemann cloned a salamander by convincing a single embryo to split into two; Briggs and King went a step further. They worked with frogs and found that they could move the nucleus from one cell to another by sucking it up into a fine glass tube. This is an important step in cloning. It led to the 1962 discovery by John Gurdon of Oxford University that the nucleus from a differentiated cell can go back to basics and build a whole organism from scratch.

In essence, the method used to create Dolly in 1997 was the same as that used by these early pioneers. Ian Wilmut and his colleagues at the Roslin Institute in Scotland got an egg, removed its nucleus and persuaded it to accept one from another cell. What stunned the scientific world was the age of the transplanted nucleus that contained all the information needed to build Dolly: it was old, very old. It had already had a long life in the udder of a six-year-old ewe, when it then spent three years in a freezer waiting for its moment. This nucleus had become set in its ways, carrying out the day-to-day duties that made it part of a ewe's bosom. Then it suddenly found itself being thawed out and taking centre stage in one of the greatest experiments in the history of science. Under the spotlight of a microscope, this elderly nucleus was set a massive challenge. No longer was it enough to be part of a skin cell; it must act as if it were young again and build an entire sheep from scratch.

Given such a demand, most nuclei get terribly confused; they are unable to erase the memory of their previous life. Eventually, they throw up the arms of their chromosomes and give up the will to live. Not this one. With a bit of coaxing and a bit of luck, the udder nucleus began to behave as if it was a whippersnapper again. It began to divide and divide, giving rise to all the cells that would eventually make up Dolly's bones, her muscles and her mind.

This technique pioneered by Wilmut and his team could have many different uses, some of which raise more ethical hackles than others. For many scientists, it threw a whole new light on the future of medicine. In theory, a cloned cell, with its infinite potential, could be used to treat any number of injuries, conditions and diseases: it could yield new skin for burn victims, give strength to shattered bone, boost the immune system and repair brain damage. Unfortunately, the promise of such 'therapeutic cloning' has largely been over-shadowed by concern over 'reproductive cloning', the fear that people will make copies of themselves. Dolly has led us, some argue, down a slippery slope towards human cloning.

Wilmut has remained steadfast. There is a moral imperative, he argues, to pursue research into therapeutic cloning and all efforts to copy humans are far beyond the boundary of moral acceptability. Somewhere in between these two extremes is the idea of using reproductive cloning technology to improve the lot of endangered species.

●

For one man, this is more than just an idea. Oliver Ryder is a senior scientist at the Center for Reproduction of Endangered

Species at San Diego Zoo. He helps to run the 'Frozen Zoo' – freezers stacked with cells from more than 7000 endangered species. The collection, mainly of sperm, eggs and embryos, has been growing steadily since 1975; cells are thawed out regularly and used to boost the genetic diversity of critically endangered populations. Shortly after the birth of Dolly the Sheep, Ryder made the case for cloning as a conservation tool. 'It has the potential of minimizing loss of genetic resources as well as rescuing already lost genetic material', he wrote in an article in *Zoo Biology*, co-authored by fellow San Diego scientist Kurt Benirschke. They were careful to outline the limitations of conservation cloning. It cannot replace protecting wild animals and it is unlikely it could ever be used to resurrect extinct species, they noted.

Nonetheless, within a year of Dolly's birth a few scientists entertained this *Jurassic Park* scenario. In the 1993 block-buster movie, entrepreneur John Hammond (played by Richard Attenborough) brings the likes of *Tyrannosaurus* and *Velociraptor* to life from dinosaur DNA preserved in amber-encased mosquitoes. Dolly's arrival inspired some to think about delving into the dusty museum archives to extract and then clone DNA from recently extinct species like the dodo. It might even be possible, they mused, to get DNA from long-gone species like the woolly mammoth, provided that its cells had been well frozen in the icy wastes of Siberia. If good-quality DNA could be extracted from such specimens, they argued, Dolly-like technology could see the return of these prehistoric beasts.

In the same vein, might it be possible to breathe life into the DNA from the museum specimens that we know came from Pinta? A clone of the one-eyed female that Rollo Beck collected back in 1901 might be of particular interest to Lonesome George. It should also be possible to exploit the temperature-dependent sex determination of reptiles to create female clones from the males collected by Cookson in 1875 and the California Academy in 1906. Implausible as such a scheme might sound, could it, in theory at least, yield enough genetic variation to restore a viable population of tortoises to Pinta?

One of the first extinct species to be offered clonal resurrection was the New Zealand huia, a bluish-blackish bird with bright orange wattles draped around its mouth. It is considered sacred by the Maori, but demand from European and North American collectors had driven it from the planet by the 1920s. Professor Diana Hill at Otago University was among the scientists who began to talk of cloning the huia, describing the initiative as 'flagship research' and 'exciting leading-edge science of international significance' in a story for CNN.

Similarly, great media attention focused on another antipodean icon, the thylacine, also variously known as the Tasmanian tiger and the Tasmanian wolf. Talk of cloning the thylacine was pushed by two Sydney institutions in 2000, who argued that it might be possible to clone the beast using DNA from a pup pickled in alcohol in 1866.

And the woolly mammoth? Although the last mammoth became extinct less than 10,000 years ago, none of the finds unearthed so far has yielded much DNA. This seems to be because the specimens have repeatedly thawed and frozen in the intervening years, says Larry Agenbroad, a regular-sized mammoth biologist based in South Dakota. With all the animals that are out there, the conditions must have been right to preserve the DNA of at least one, he says. 'If we're lucky enough we'll find it.'

The best-preserved specimen to date was discovered in November 2002 by two school students in the remote Sakha Republic of northeast Siberia. The Yukagir mammoth, named after a nearby town, was carefully exhumed in June 2004. At the time, there was considerable excitement that this specimen might not have been subject to freeze–thaw cycles that destroyed the DNA in all previous mammoths. 'The condition of the skull and left front leg are optimum', explains Agenbroad. He was right. When Japanese geneticists got their hands on the remains of this 18,000-year-old mammoth in 2005, they found the DNA well preserved, obtaining 'the first complete sequence of the mitochondrial genome from an extinct organism of prehistoric age'. Apparently it is still not enough to create a clone.

As long as a plant or an animal is alive, a host of enzymes is busy repairing any DNA damage that occurs. As soon as it is dead, another bunch of enzymes get the upper hand and begins to pull the DNA apart. In the movie version of *Jurassic Park*, Michael Crichton acknowledged this problem and invented an ingenious solution. 'It's full of holes', announces an on-screen cartoon strand of DNA that explains the bogus science behind dinosaur cloning to visitors. 'That's where our geneticists take over', it babbles, gyrating its helical body. 'Thinking machines, super computers and gene sequencers break down the strand in minutes, and virtual reality displays show our geneticists the gaps in the DNA sequence. We use the complete DNA of a frog to fill in the holes and complete the code. And now we can make a baby dinosaur.' Pure and delightful science fiction.

It might, one day, be possible to repair DNA damage, says Alan Cooper, an ancient biomolecules expert from the University of Adelaide in Australia. But samples from a long-dead specimen like the Yukagir mammoth or even a relatively recent casualty like the dodo have accumulated far too much damage to be of any use to conservation, he says. Filling in the gaps that the Yale geneticists found in the DNA from century-old tortoise specimens is more fiction than science, he cautions. Lonesome George will never be pairing off with clones of the museum-based Pinta animals.

Whilst he is still alive and his DNA is still in one piece, what about a spot of icon cloning to capture his special set of genes? Soon after Ryder and Benirschke's *Zoo Biology* article, reports began to emerge of efforts to clone endangered species. In March 1999, scientists at Utah State University published a paper detailing their attempt on an argali – a wild sheep from the mountains of central Asia considered vulnerable to extinction. They got a few pregnancies, but the foetal argali clones died long before they were born.

Then in early 2001, scientists at Advanced Cell Technology (ACT), a private company in the US, announced that they had cloned a gaur, a wild ox they described as 'on the verge of extinction'. Notwithstanding the 30,000 still roaming around central and Southeast Asia, the World Conservation Union is concerned enough about ongoing population decline that the gaur is on its list of threatened species. In October 2000, with the baby gaur still in the womb, ACT researchers knew that they were just weeks away from being the first to produce an endangered species clone. 'This study presents exciting possibilities for those of us working to preserve endangered species', said Robert Lanza, vice-president of medical and scientific development at ACT. When the baby gaur was born by Caesarean section on 8 January, they were bold enough to name him Noah. Two days later, he was dead, reputedly from a nasty bacterial infection.

That same year an Italian team fared better in cloning a mouflon, a mountain sheep found only on the Mediterranean islands of Sardinia and Corsica. The baby clone came about by chance, says Pasqualino Loi of the University of Teramo in Italy. Scientists at a wildlife rescue centre in Sardinia had been custodians of a small herd of elderly and injured mouflon, when two females died unexpectedly. They were found lying in the pasture, Loi recalls. This was a bit of a setback for the institution: one of the females – known as Ombra – had been the most fertile female in the captive herd. Ever resourceful, it occurred to Loi that eggs from Ombra's ovaries might be harvested and fertilized in the lab with sperm donated by a willing mouflon male. In theory, the eggs could be used to create hundreds of test-tube mouflon, so Ombra's inadvertent death might turn out to be a boon for a Sardinian herd estimated at only around 2500 individuals.

The team harvested ovary cells from Ombra but could not get them to divide and release eggs. Loi was on the brink of giving up when another idea came to him. The nuclei within Ombra's cells might still be working normally. Just as the nucleus from an udder cell was the blueprint for Dolly, so a nucleus from one of these dead females might make a mouflon clone.

Although the argali, the gaur and the mouflon are all considered endangered, there is a good reason why biologists do not try to clone more charismatic species like the Javan rhino, the Siberian tiger or even the Pinta tortoise. Cloning is only occasionally hit and much more often miss. Out of 277 cloned sheep cells that Wilmut and his Roslin team created, only 29 began to divide normally. These were put into 13 surrogate ewes, of which only one fell pregnant. So behind the singular, if significant, success that is Dolly, there is a level of failure that sits uncomfortably with conservation aims. Creating a clone requires hundreds of female animals to supply eggs and act as surrogate mothers for the few embryos that begin to divide. Nowhere in captivity and probably not even in the wild is there a large enough band of female Siberian tigers; Javan rhinos are even rarer; and you can't get more peerless than the Pinta tortoise.

By contrast, the argali, the gaur and the mouflon are so similar to farmyard animals that these common cousins were recruited to do all the hard work for their endangered relatives. Ombretta came about by transferring a mouflon nucleus into an egg from a baa-standard domestic sheep. The fancy nucleus took surprisingly well to its humdrum home – nearly a third of all embryonic clones began to divide. Each of these was introduced into a sheep, resulting in two pregnancies, one of which reached term. So one species, a sheep, gave birth to another, a mouflon. The infant was called Ombretta, Italian for 'little shadow'. She lived for six months.

Figure 11.1 Ombretta, the mouflon clone

Biotech coup or not, some conservationists are suspicious of creations like Ombretta. While she had the nuclei of a mouflon, looked like a mouflon and walked like a mouflon, she still had something of the sheep about her: the rest of each cell. The jury is still out on the consequences of mixing cells from different species.

For Lonesome George's genes, there seems to be no alternative. If he is to be cloned, a ready supply of eggs will be needed. Because there's no Pinta female to provide them, the eggs will have to come from a female from a different island. Just as Ombretta was not a perfect mouflon copy, Lonesome George's hypothetical cloned descendants could never be pure Pinta. Such a compromise is, perhaps, the only way out of the evolutionary cul-de-sac that Lonesome George's genes now find themselves in.

Cloning a reptile is still a very long way off. This is partly down to lack of cash: reptiles are not hugely commercial creatures, so there's little money available for the basic research needed. It's also because of 'the shell problem'.

Up to a point, the recipe for cloning egg-laying animals such as tortoises reads just like that for mammals. Take one egg; suck out and bin its nucleus; choose a cell from the animal you wish to clone (one of Lonesome George's muscle cells, say); remove and keep its nucleus; insert this nucleus into the eviscerated egg. Give the egg-with-new-nucleus combo a jolt of voltage. Hey presto, you have your cloned cell. This is where the similarity ends.

Just as a reptile egg fertilized in the laboratory by IVF or ICSI needs a yolk and a shell to thrive, so too would a cloned embryo. 'The shell problem', probably more than anything else, is what will really thwart anyone wanting to use lab-based technologies to help out Lonesome George. But since

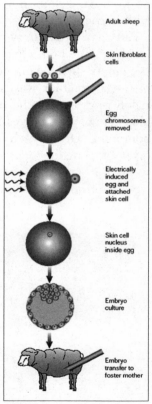

Figure 11.2 Cloning, in six easy steps

we've come this far, hell, let's explore a couple of possible solutions to this conundrum.

With a bit of cunning, it might just be possible to do away with the shell and yolk altogether, and nurture a test-tube tortoise in the laboratory all the way from a single cell to a hatchling. As long ago as 1988, Margaret Perry at the Roslin Institute managed to 'grow' a chicken in the laboratory, nurturing a ball of cells into a ball of fluff in around three weeks. She started off with a fertilized egg, sealing it in a glass jar for a day before moving it into a surrogate eggshell filled with nutrients.

Perry made the surrogate shells from chicken eggs by slicing off the pointed end, tipping out the yolk and white, washing the empty shell clean and filling it with nutrients. Once the microscopic chick was safely inside, she used clingfilm and nylon rings to seal the hole in the top of the shell. As the chick grew, Perry moved it again into another surrogate shell containing a fresh batch of nutrients, this time gluing a double layer of clingfilm over the hole in the top. Incredibly, it worked. In principle, there's no reason why a similar method would not work for an embryo cloned from Lonesome George.

Figure 11.3 Perry's chick inspects its surrogate eggshell

It would be much easier, though, to hijack another egg, than to go to all the trouble of making your own eggshells and nutrients to feed your cloned embryo. In the poultry industry, it is common practice to inject chemicals into eggs to protect chicks against disease or supplement their early growth. There's nothing to stop this technology being adapted to allow the injection of cloned cells into an existing embryo.

Although this is less labour-intensive than Perry's surrogate eggshell method, it does have a drawback. The hatchling that emerges will not be a pure clone. It will be a combination of two individuals, with some of its cells coming from the original embryo and some coming from the injected clone. It will be a chimera.

In case you're wondering, there is a subtle but important difference between a hybrid and a chimera. If Lonesome George and an Española female ever managed to reproduce naturally, their offspring would be a Pinta-Española hybrid and each cell would contain a blend of Pinta-type and Española-type genes. A Pinta-Española chimera, by contrast, would have two sorts of cells in its body – some of Pinta origin and some of Española origin. Think of it this way. You are given two lumps of clay – one red and one yellow – and asked to mould them into the shape of a tortoise. Blending your clay into one ball will make an orange hybrid tortoise; sticking yellow legs and head onto a red shell will give you a chimera.

The ancient Greeks came up with the idea of the chimera, imagining creatures such as the centaur (a horse with a human head) and the griffin (a lion with an eagle's wings). In the 21st century, geneticists have manipulated embryos to give rise to chimeras like the 'geep' (part goat, part sheep). The same approach might keep certain precious genetic combinations alive, and could be particularly useful for egg-laying creatures like birds and reptiles. For example, cells cloned from Lonesome George could be injected into the embryo of an Española (or other) tortoise. The resulting baby tortoise would be a chimera – a mixture of cells from the Española host and cells cloned from Lonesome George. At least his unique set of DNA would still be working away inside a living tortoise rather than being destined for the evolutionary scrap heap.

Taking this idea one step further, it should be possible to direct the injection in very specific ways. In 2004, a Japanese team engineered salmon to produce trout sperm and eggs. They made their fishy chimera by taking cells from a trout and injecting them into the ovaries or testes of young salmon. Consequently, some of the salmon grew up ejaculating trout sperm and some grew up laying trout eggs. So if cells cloned from Lonesome George were injected into an Española embryo, there's a tiny chance that some would end up in the testes or ovaries. If it could be pulled off, this Pinta-Española chimera might start making sperm or eggs containing just Lonesome George-type genes.

Clearly, such ideas are armchair science. A huge amount of research into the basic embryology of reptiles is needed even to begin thinking about this kind of intervention. This means money, and if there's one thing conservation biologists agree on, it's that precious funds would be better spent elsewhere than on trying to clone a reptile. What if money *were* available for research to clone Lonesome George?

The history of science is littered with stories of benevolent patrons (occasionally matrons) funding some obscure bit of research that simply would not have been done were it not for their rather esoteric interests. A cracking example is Donald Jacklin, a businessman and mule-racing enthusiast in Idaho, who ploughed some of his wealth into cloning a mule.

Mules are the hybrids that result when a horse and a donkey

get it together. Because mules are sterile, this has helped to avoid strong opposition to the project. In 2003, three genetically identical mules – Idaho Gem, Utah Pioneer and Idaho Star – were born of the investment, and the media-savvy Jacklin won over the hearts and minds of a global audience. In 2004, the three mules were the star attraction at the annual meeting of the American Society for the Advancement of Science in Seattle. They are pretty comfortable with all the attention, says Gordon Woods, professor of animal and veterinary science at the University of Idaho and one of those who benefits from Jacklin's largesse. 'Once the cameras start clicking, they start performing', he says. Idaho Gem and Idaho Star are destined for a life as racing mules.

Figure 11.4 Idaho Gem, the mule clone

So some wealthy tortoise enthusiast, and there are plenty of them, might one day step forward with a bundle of cash on the understanding that it goes towards research to clone Lonesome George. It would be silly to turn down such an offer.

But what would it actually achieve? What point could there be in cloning this shy, retiring and enigmatic creature? Even if it were possible to create both girl and boy Georges by manipulating incubation temperature, putting together a group of genetically identical individuals can hardly be considered a population. What's more, any offspring produced by these clones would be very likely to have all the health and fertility problems that characterize highly inbred individuals.

Ultimately, the future of the Pinta race rests on finding another Pinta tortoise that can bring just a smidgen of genetic diversity to the table. If this hypothetical tortoise turns out to be a female, great. But another Pinta male would be good news too. If cloning were an option, the baby clones from this male could be incubated warm to make females, giving Lonesome George clones a mate with different genes and the vague hope of producing some healthy offspring.

When Aldous Huxley wrote *Brave New World* in 1931, he set his fiction in the seventh century AF (after Ford). Just a quarter of a century after its publication the following year, he was stunned to see his fiction becoming reality. 'The prophecies I made in 1931 are coming true much sooner than I thought they would', he wrote in *Brave New World Revisited*, published in 1958.

There are, as yet, no cells from Lonesome George in a freezer; nobody has been able to crack the conditions needed to cryopreserve a sliver of giant tortoise. Hopefully, someone will work out how before too long. Then, even if Lonesome George is with us for another century (and let's hope so), his special set of genes will outlive even him. The day may yet come when there is a really good reason to thaw them out.

Epilogue
WHAT NOW?

Whoever named Lonesome George got it spot on. We can be virtually certain that there are no tortoises still to be discovered on Pinta and that the World Conservation Union is right to describe Lonesome George's subspecies *Geochelone nigra abingdoni* as 'Extinct in the Wild'. It is unlikely that ongoing searches of the world's zoos and private collections will uncover a giant tortoise with Pinta-type genes and look. There's a good chance that this despondent fellow is indeed one of a kind.

When Española tortoises are set loose on Pinta, as now seems probable, should Lonesome George join them? Plenty of people opposed to keeping wild animals in captivity will voice support for his repatriation. There is some merit in the argument that keeping George (and all other captive animals) confined by stony walls for the rest of his life is inhumane and inappropriate. Surely it would be fairer to George to set him free on his native soil and let him roam its misty, muggy wilderness until his dying day?

Not necessarily. Pinta is a treacherous place, particularly for giant tortoises. The plethora of bones and shells recovered from the deep fissures that score the island is testimony to this. If we had never discovered Lonesome George, he probably would have met this same end sooner or later. If he is returned there one day, this is almost certainly the fate that awaits him. Consigning a unique animal to death in a gloomy crag, suffering from a seeping crack to his shell and creeping hunger would certainly be inhumane. More importantly, perhaps, it would be a wasted opportunity.

Figure 12.1 Lonesome George at home on Pinta in 1972

George, as I'm wont to repeat, is a conservation icon. The voice he offers endangered species is loud and clear. Like it or not, his celebrity status bestows upon us and him an obligation to the rest of the natural world. He must remain in captivity for the rest of his years and continue the good work he has started. As a flagship, he must keep on preaching the conservation message and converting tourists to the conservation cause. And his custodians at the CDRS must give him every opportunity to do so.

Getting George to reproduce is, frankly, a long shot. This doesn't mean that we shouldn't try. Everything that can reasonably be done should be done; every initiative, every attempt enriches and strengthens his story and his power to communicate.

Linda Cayot's effort to get him to hybridize with the Isabela females was just the ticket. Now that geneticists have revealed his closest relatives come from Española, it is about time the Isabela tortoises were taken off harem duty and replaced by some new, young blood from the immensely successful Española breeding programme. And while tortoises are being moved about the research station, why not give George more opportunity to interact with others, particularly males, to get him revved up a bit.

Sveva Grigioni's stint at the CDRS is downplayed by some scientists for fear of ridicule. It shouldn't be. It's the closest that anyone's come to getting George into sexual shape. It should be tried again. If nothing else, it will improve his mood. At the same time, it would be little trouble to take a swab from his cloaca or check his faeces for sperm. If there are any there that can't be extracted by manual means, it might be worth thinking hard about interventions like electroejaculation.

Whatever happens, there's no reason why Lonesome George should not outlive us all, acting as a focus for Galápagos conservation for generations to come. I hope so. I like the thought that when my son grows up, I will take him to the Galápagos and introduce him to George.

It all depends on what happens in the archipelago in the next few years. The pace of social change is alarming and does not bode well for the future of the islands' unique suite of species. Let's hope the Special Law can tame the population expansion. Tourism in the Galápagos Islands currently generates at least $150 million a year. If more of this can be fed back into a sustainable future, then yes my son may yet see the same pristine, enchanting world I saw. That Darwin saw.

One day, of course, George will give up the tortoise ghost. Even then, he will be of immense value to the Galápagos. His remains should not be taken back to his native island. Nor should they be flown to Quito to act as a centrepiece in the Museo Ecuatoriano de Ciencias Naturales. Lonesome George must remain in the archipelago, at the research station on Santa Cruz. By then, this is where he will have spent most of his life; this is the place that Lonesome George would call home. Even in death, it is here that he will have his greatest audience.

I, for one, would like to see him again – alive or dead. I feel I'm just about getting to know this tortoise, and I like him.

NOTES AND SOURCES

This is a chapter-by-chapter account of the sources used to furnish
Lonesome George *with historical and scientific detail and quotations. Full*
references are given in the Bibliography and further reading that follows.

Prologue – A conservation icon

'I had the feeling …': Sveva Grigioni, personal communication
'The natural history of these islands is eminently curious': Darwin (1845)

Chapter 1 – Discovery

'The tortoise was walking slowly …': Vagvolgyi (1974)
The Vagvolgyi's dinner party: Peter Pritchard, personal communication;
  see also Pritchard (1977)
'The tortoise we saw on Pinta …': paraphrased from Pritchard, personal
  communication
'I practically lost my teeth': Pritchard, personal communication
'I was a pretty weird kid': Pritchard, personal communication
'a peculiar beast …': Pritchard (1977)
'I resolved …': Pritchard (1977)
Rediscovery of the woolly flying squirrel: Peter Zahler, personal
  communication
Rediscovery of the Santiago rice rat: Robert Dowler, personal
  communication; see also Dowler et al. (2000)
'It's a woolly flying squirrel': Zahler, personal communication
'I remember seeing bleached bones …': Peter Kramer, personal
  communication
Details of the goat-hunting trip to Pinta in March 1972: Manuel Cruz,
  personal communication; see also Cruz (1994)
'We both thought it was a goat …': Cruz (1994)
'The swinging of the tortoise …': Cruz (1994)
Details of Ole Hamann and Peter Pritchard's trip to Pinta: Ole Hamann
  and Peter Pritchard, personal communication; see also Pritchard (1977)
'I was much more excited …': Pritchard, personal communication
'They have found a tortoise, a large male': paraphrased from Ole Hamann,
  personal communication
'There was still hope …': Pritchard (1977)
'It lacked the antediluvian look …': Pritchard (1977)
George's movements on arrival at the CDRS: Kramer, personal
  communication

Status of tortoise populations throughout the archipelago: see MacFarland
(1974a)
Pinzón egg collection: see Perry (1970); MacFarland (1974b)
Captive breeding of Española tortoises: see Perry (1970); MacFarland
(1974b)
Details of Pritchard's week on Pinta: Pritchard, personal communication
'The utter senselessness of it …': Pritchard (1977)
Arrival of Vagvolgyi's photograph: Pritchard, personal communication
Pritchard and Kramer compare Vagvolgyi's photograph to Lonesome
George: Pritchard and Kramer, personal communication
'George is about as short and inconspicuous-looking …': Anon. (1954)
Gobel bowls 'em over, *TV Guide* 2, 5–7
Counting rings works for young tortoises: Kristin Berry and Linda Cayot,
personal communication
Lonesome George's age: Cayot and Pritchard, personal communication
Marion's tortoise: Gerlach (1998)
Tui Malilia: see www.guinnessworldrecords.com; Gerlach (1998)
Harriet's dubious origins: Chambers (2004a)
'Harriet's DNA shows signficant differences …': Chambers (2004b)
George's fall: Anon. (1980) Lonesome George achieves immortality – in
bronze. *Noticias de Galápagos* 32: 2–3
George's weight gain: Cayot, personal communication
George's neck swelling: Cayot, personal communication
George's constipation: Joe Flanagan, personal communication
Treatment by nutritionist: Olav Oftedal, personal communication

Chapter 2 – Lonesome George's girlfriend

Overview of George's life at the CDRS: Linda Cayot and Roslyn Cameron,
personal communication
Outbreeding depression in ibex: cited in Frankham et al. (2002); van
Wieren et al. (2005)
Hybridization of red wolves with coyotes: see Frankham et al. (2002);
Wayne (1996); Wilson et al. (2000)
Inbreeding depression in the Florida panther: see Culver et al. (2000)
Townsend visit to collect the last remaining Galápagos tortoises: Townsend
(1928)
Efforts to breed tortoises at the CDRS: Corley Smith (1976)
'Visit Lonesome George, the world's oldest living gay turtle':
www.flamingo-travel.com
Ubiquity of homosexuality in the animal kingdom: Bagemihl (2000)
Early learning influences adult behaviour: see Freeberg (2000)
Japanese quails watching TV: Ophir and Galef (2003)
'Some animal species show increased interest …': Fritts (2002)
'Potentially the visual, olfactory, and even auditory stimuli …': Fritts
(2002)

Captive breeding of Seychelles and Arnold's tortoises: Justin Gerlach, personal communication

von Hegel's technique: Gisela von Hegel, personal communication

'She could do this in just a few minutes': Cayot, personal communication

Semen collection and artificial insemination in elephants: Thomas Hildebrandt, personal communication

'looks like the greatest Swiss cheese': Hildebrandt, personal communication

'the penis is actually a structure that you never should touch': Hildebrandt, personal communication

Details of Sveva Grigioni's work with George: Grigioni (1993); Grigioni and Cayot, personal communication

'Sveva could get the other male tortoises …': Cayot, personal communication

'He was very shy at the beginning': Grigioni, personal communication

'if a single female moth were to release all the bombykol …': Lewis (1995)

Wedekind's T-shirt experiment: Wedekind et al. (1995)

Mexican rams: Lezama et al. (2001)

'Day by day, he started to be more interested in the females …': Grigioni, personal communication

'If I had had more time …': Grigioni, personal communication

Chapter 3 – The origin of a species

George's feeding routine: Roslyn Cameron, personal communication; Joe Flanagan, personal communication; see also 'The story of Lonesome George': www.darwinfoundation.org/Restoring/george.html

'such big tortoises that each could carry a man on top of himself': de Berlanga (1535)

Debate surrounding tortoise origins: Larson (2001)

Beebe's experiments with the Duncan Island tortoise: Beebe (1923, 1924)

'In spite of frequent slipping it kept obstinately ahead …': Beebe (1924)

'After we had handled it a few times …': Beebe (1924)

'When placed in the water alongside the ship …': Beebe (1923)

'I could see the throat vibrate in breathing …': Beebe (1924)

'A week later this tortoise died without warning': Beebe (1924)

Albert Günther's early ideas about tortoise origins: Günther (1875)

George Baur's thinking on tortoise origins: Baur (1889, 1890, 1891a, 1891b)

'The Galápagos originated through subsidence …': Baur (1891a)

'All hands are anxious to leave as soon as possible …': see Fritts and Fritts (1982)

'They would occasionally stick their heads out of water …': see Fritts and Fritts (1982)

California Academy lose and then recover two tortoises from the ocean:
    see Fritts and Fritts (1982)
'When they drift on island shores …': Van Denburgh (1914)
'We must rather adopt the view that the islands …': Van Denburgh (1914)
'may arrive, after a passage of several weeks, at the bay of an island …':
    Lyell (1830)
Colnett's observation of driftwood: Colnett (1798)
'driftwood, not the growth of these islands …': Fitzroy (1839)
'the seeds of 14/100 plants of any country …': Darwin (1859)
'I do not deny that there are many and grave difficulties …': Darwin
    (1859)
'I have not found a single instance …': Darwin (1859)
Mayotte island frogs: Vences et al. (2003)
'I am in perfect agreement with Van Denburgh…': Beebe (1924)
'It is pertinent to inquire, why was it so little used?': Townsend (1925a)
'it is inconceivable that various forms of living flotsam …': Townsend
    (1925a)
'The ancestry of the island tortoises …': Townsend (1925a)
Yale analysis of tortoise origins: Caccone et al. (1999); Adalgisia Caccone,
    personal communication
Dodo weight: Maddox (1993)
Elephant bird size: Line (1994)
Australian tiger snakes: Keogh et al. (2005); Scott Keogh, personal
    communication
Haast's eagle: Bunce et al. (2005)
Goliath the giant tortoise: Gerlach (1998); Greg Moss, personal
    communication
The case for floating giant tortoises: Pritchard (1979)
Timeframe of giant tortoise evolution: Caccone et al. (1999); Beheregaray
    et al. (2004)
Drowned islands: Christie et al. (1992)
Iguana evolution: Wyles and Sarich (1984); Rassmann (1997)

Chapter 4 – Random drift

'There is every reason for believing …': Darwin (1839)
Erroneous classification of tuatara: Daugherty et al. (1990)
Galápagos giant tortoise taxonomy: Pritchard (1996)
Yale analysis of Santa Cruz tortoise genetics: Russello et al. (2005)
'In space and time …': Darwin (1845)
Darwin's path through the archipelago: Estes et al. (2000)
The significance of the Galápagos to Darwin: Frank Sulloway, personal
    communication; Sulloway (1982, 1984, 1985)
'His primary interest was in the geology …': Sulloway, personal
    communication

'I frequently got on their backs …': Darwin (1845)

'One large one, I found by pacing …': Keynes (2000)

'It was greatly astonished': Darwin (1845)

'I several times caught this same lizard …': Darwin (1845)

'It would appear that the birds of this archipelago …': Darwin (1845)

'In my walk I met two very large Tortoises …': Keynes (2000)

'One was eating a Cactus …': Barlow (1934)

'It was confidently asserted, that the tortoises coming from different islands …': Darwin (1839)

'I have not as yet noticed by far the most remarkable feature …': Darwin (1845)

'So it is at least possible that tortoises …': Thornton (1971)

When extremely heavy rain begins to fall …': Linda Cayot, personal communication

'If those were the only two tortoises …': Sulloway, personal communication

Tortoise morphology: Fritts (1983, 1984)

Migration of tortoises throughout the archipelago: Caccone et al. (2002); Ciofi et al. (2002); Beheregaray et al. (2004); Caccone, personal communication

Isabela aliens: Caccone et al. (2002); Caccone, personal communication

USS *Essex* capturing the *Georgiana* and *Policy*: Porter (1813, 1815); Farragut (1879)

'Sail ho! Sail ho!': Porter (1815)

'At eleven A.M., according to my expectation …': Porter (1815)

'In clearing their decks for action …': Farragut (1879)

'At two o'clock, the boats were about a mile …': Porter (1815)

'rowed up beneath the muzzles of the guns …': Porter (1813)

'Thus were two fine British ships …': Porter (1813)

'they had been lying in the same place …': Porter (1815)

'and had supplied themselves abundantly …': Porter (1815)

Cerro Montura tortoises: Russello et al. (2005)

Cookson's tortoise observations: Cookson (1876a, 1876b)

'As Hood and Abingdon Islands …': Cookson (1876b)

'There is a strong current running northwest …': Caccone et al. (1999)

Lava lizard parallel: Wright (1983, 1984); Kizirian et al. (2004); David Kizirian, personal communication

Edward Louis' analysis of tortoise genetics: Edward Louis, personal communication

Yale analysis of ancient DNA from California Academy tortoises: Caccone et al. (1999)

Chapter 5 – Man trap

Flagship species: see Leader-Williams and Dublin (2000); Walpole and Leader-Williams (2002); Caro et al. (2004)

'Whatever happens to this single animal …': Charles Darwin Research Station (2002)

Avocet as a flagship: see www.rspb.org.uk

Hsing-Hsing and Ling-Ling: Anon. (1999) National Zoo's giant panda Hsing-Hsing dies, CNN Washington; Anon. (2000) Giant pandas: an epic tale at the National Zoo, *Communiqué,* February

'They represented the world of nature …': Anon. (2000)

Chi-Chi: www.wwf.org/

Elsa: see www.bornfree.org.uk

Brighty the Grand Canyon donkey: Wills, J. (2005) *Donkey Politics in the Grand Canyon*; Annual meeting of the British Society for the History of Science, University of Leeds, 15–17 July; John Wills, personal communication

'Brighty provided a way to negotiate the landscape': Wills, J. (2005) *Donkey Politics in the Grand Canyon*

Komodo dragon as a flagship: Walpole et al. (2001)

'A group rather of extinct volcanoes than of isles …': Melville (1854)

'The appearance of this man …': Porter (1815)

Orchil industry in 19th century: see Hickman (1991)

Development of tourism in the islands: see de Groot (1983); MacFarland (1998)

American Acclimatization Society: Ingold (1989)

'Nay, I'll have a starling …': Shakespeare, *Henry IV*

Rabbits: see www.csiro.au/

Cane toad: see www.csiro.au/

Red fire ant: see Moloney and Vanderwoude (2002)

Black rats: Donna Harris, personal communication

Philornis downsi: Fessl and Tebbich (2002)

Red quinine tree: Buddenhagen et al. (2004)

Control of cottony-cushion scale using the ladybug beetle: Causton et al. (2004); Causton (2005)

'The mangrove stands of Puerto Ayora …': Causton (2005)

'He had already procured a little heap …': Darwin (1845)

Response of marine iguanas to tourists: Romero and Wikelski (2002)

Development of the archipelago: MacFarland and Cifuentes (1996); Snell et al. (2002); Anon. (2003) Energy evolution: renewable energy in the Galápagos Islands, *Refocus* **4**(5): 36–8; Ley (2003); Kerr et al. (2004); Kerr (2005); Boersma et al. (2005); White (2005); Linda Cayot, personal communication; Graham Watkins, personal communication; Leonor Stjepic, personal communication

Jessica spill: Galápagos National Park press releases; see www.galapagos.to/texts/jessica.htm

'All of us, as well as the horses …': de Berlanga (1535)

Woram's analysis of Berlanga's letter: Woram (2005)

Special Law for the Galápagos: see Bensted-Smith (1998); Heylings (1999)

Chapter 6 – Lock up your tortoise

Sea cucumber biology: Chantal Conand, personal communication;
Verónica Toral-Granda, personal communication
Sea cucumber fishery: MacFarland and Cifuentes (1996); Okey et al.
(2004); Shepherd et al. (2004); Altamirano et al. (2005); Hearn et al.
(2005); Toral-Granda and Martinez (2005); Toral-Granda (2005);
Graham Edgar, personal communication; Toral-Granda, personal
communication
'If I had been born in the Guasmo of Guayaquil …': Merlen (1993)
'As the islanders saw that a three-man crew …': D'Orso (2003)
'It is a classic case of a group of relatively poor people …': Powell and Gibbs
(1995)
'It started with them just sitting on the front steps …': Linda Cayot,
personal communication
'Hundreds, and perhaps thousands, of Ecuadorians …': Valle (1994)
Isabela fire: Marquez et al. (1994); Novak (1994); Powell and Gibbs (1995)
Tortoise killings: Cayot and Lewis (1994); Cayot, personal communication;
Jim Pinson, see www.jc-research.com/jim/
'The destruction caused by the fire …': Wikelski (1995)
Details of unrest during 1995: Chantal Blanton, Verónica Toral-Granda
and Linda Cayot, personal communication; Jim Pinson's archived
emails, see www.jc-research.com/jim/
'Certainly sea cucumber fishermen …': Cayot, personal communication
'sustainable use of the natural resources by the local population': Anon.
(1995) Insurrection in Galápagos, *Galapagos Conservation Trust
Newsletter*, autumn
'No country in the world allows treasures …': Barry (1995)
'We are prepared to take extreme steps …': Velíz (1995)
'They had come to take him by force …': Pinson's archived emails
'We were asked to leave all personal belongings': Toral-Granda, personal
communication
'We had to make sure the baby iguanas …': Toral-Granda, personal
communication
'It allows us to begin discussions about sustainability …': Watkins, personal
communication
Ivory markets in Europe: Martin and Stiles (2005)
Sariska Tiger Reserve: Narain et al. (2005)
Elephant–human conflicts: Osborn and Parker (2002, 2003)
'Elephants hate chilli': see www.elephantpepper.com/
Ethiopian wolf: Claudio Sillero, personal communication; see
www.ethiopianwolf.org
Inferno Community Ecotourism Project: Stronza (2000)
Direct payment for conservation: Balmford and Whitten (2003); Ferraro
and Kiss (2002); Ferraro and Simpson (2003)

Direct payment for Costa Rican forests: see Malavasi and Kellenberg (2002)
Tragedy of the commons: Hardin (1968)
Guadalupe River confiscations in 1997: Vargas (1997)
Belle Vie shooting on 19 March 1997: Cruz (1997a, 1997b)
San Cristóbal raid on 28 December 2003: Anon. (2003) 15,933 Sea Cucumbers found in San Cristóbal, *Galapagos Conservation Trust*, www.gct.org
Cargo ship in March 2004: Anon. (2004) In Galápagos the trade in sea cucumbers goes on, *Galapagos Conservation Trust*, www.gct.org
Political instability: see Anon. (2004) Quick exit for Galápagos director. *Science*, **306**(5694), 225

Chapter 7 – The mysteries of Pinta

'Yonder, though, to the E.N.E. ...': Melville (1854)
'Friedrich found himself involved in a protracted private war ...': Treherne (2002)
Pig control on Santiago: Cruz et al. (2005)
Impact of goats on Pinta: Hamann (1993)
'Extinct in the Wild': Tortoise & Freshwater Turtle Specialist Group (1996).
Cebu flowerpecker: Magsalay et al. (1995)
Ivory-billed woodpecker: Fitzpatrick et al. (2005)
'We, with 7 of our men, went to discover this island ...': Cowley (1687–89?)
'We were close under Abington Isle ...': Colnett (1798)
Impact of whaling industry on Galapagos tortoises: Townsend (1925a)
Details of botany expedition: Derek Green, personal communication; Rob Gradstein, personal communication
'Pinta at the time seemed to us the most unspoiled of all the islands ...': Gradstein, personal communication
'I don't think there were any species records ...': Gradstein, personal communication
'On that particular day I just had this gut feeling ...': Green, personal communication
'It hadn't been dead more than a year or so ...': Peter Pritchard, personal communication
'I can still envision the precise point ...': Linda Cayot, personal communication
'Definitely sort of old and withered': Hamann, personal communication
'Greyish-brown, with a coarse, fibrous structure...': Seberg, personal communication
El Niño rainfall: Snell and Rea (1999)
Lynx scats in Portugal: Pires and Fernandes (2003)

'It is possible to misidentify scats …': Fernandes, personal communication

Grizzly poo: Wasser et al. (2004)

Collecting dolphin faeces: Parsons et al. (1999, 2003)

'You have to be able to locate, swim to and collect the faeces …': Parsons, personal communication

Rare sightings of Santa Fe tortoises: Howard Snell, personal communication

'It had been used for the same purpose that rocks …': Fritts and Fritts (1982)

'Getting my pack, I ate supper and skinned the tortoise by moonlight': Van Denburgh (1914)

Possible G. *nigra phantastica* scats found in 1964: Perry (1970)

Shipton's exploration of Fernandina: see Corley Smith (1976)

'Using experienced tortoise searchers …': Fritts et al. (1998)

Quest aired on the Discovery Channel: Thomas Fritts, personal communication

Pritchard hatches a plan to search Pinta: Pritchard, personal communication

Details of Pinta 2003 expedition: Pritchard (2004)

'Most appeared to be of adult animals …': Pritchard (2004)

'It was intact, except for a large hole in the left side …': Pritchard (2004)

'From that, I was assuming that was an old female': Pritchard (2004)

'There are areas … that appear eminently suitable …': Pritchard (2004)

'They come down from the mountains …' Porter (1815)

Temperature-dependent sex determination: see Shine (1999)

Unusual Pinta geology: William White, personal communication

Nest site fidelity in turtles: Freedberg and Wade (2001)

Chapter 8 – The diaspora

$10,000 reward: cited in Marquez et al. (2001)

'There was a sense that this opportunity was going to end': Edward J. Larson, personal communication

Details of Commander Cookson's visit: Cookson (1876a, 1876b)

'They are still tolerably numerous near Tagus Cove': Cookson (1876b)

'One of these, owing to want of sufficient hands …': Cookson (1876b)

'to lower them over the cliff, a height of about 200 feet': Cookson (1876b)

'I could not preserve all alive …': Cookson (1876b)

California Academy expedition: Van Denburgh (1914); Fritts and Fritts (1982)

'It is capital country for tortoises': Van Denburgh (1914)

Rothschild's reaction to Cookson's letter: see Rothschild (1983)

'I believe they are doomed …': Cookson (1876b)

'to bring away every tortoise they saw …': cited in Rothschild (1983)

Details of Webster–Harris expedition: Rothschild and Hartert (1899)

'I instructed each man to collect …': cited in Rothschild and Hartert (1899)

'Do not think tortoise exist here …': cited in Rothschild and Hartert (1899)

Beck ships 'old mossback' to Rothschild in summer of 1901: Beck, R. H. (1901) Letter to Dr Ernest Hartert, dated 8 July 1901, Berryessa, California, Rothschild Collection, Natural History Museum, London

'at the foot of the cliff …': Beck to Hartert, 8 July 1901

'One day our largest tortoise …': Beck to Hartert, 8 July 1901

'If you wish these as skins …': Beck to Hartert, 8 July 1901

'If they will I will ship alive …': Beck, R. H. (1901) Letter to Dr Ernest Hartert dated 10 September 1901, Berryessa, California, Rothschild Collection

Beck skins the one-eyed Pinta female: Beck, R. H. (1901) Letter to Dr Ernest Hartert, dated 29 October 1901, Berryessa, California, Rothschild Collection

One-eyed female could offer vital clues in the search for a mate for George: Pritchard (1984)

Rediscovery of the forest owlet: Pamela Rasmussen, personal communication

'Periods of malnutrition or rapid growth …': Thomas Fritts, personal communication

Unusual shell shape of Pritchard's 1972 find: Pritchard (1984); Pritchard, personal communication

Possible Pinta tortoises at the Smithsonian: Fritts, personal communication

Yale analysis of DNA from unidentified tortoises: Burns et al. (2003)

Pinniped penises: Malik et al. (1997)

Genetic analysis of Galápagos tortoises by Edward Louis: Edward Louis, personal communication

DNA sampling of South American captive tortoises: Michael Russello, personal communication

Tony at Prague Zoo: Petr Velenský, personal communication

Spix's macaw: Juniper (2002)

Return of Presley to Brazil: Anon. (2003) Spix's macaw returned to sender, *Birdlife International*, 14 April 2003

Caloosahatchee Aviary: Greg Moss, personal communication

Chapter 9 – Wild at heart

'an attempt to establish a species …': IUCN/SSC (1998)

Arabian oryx: Spalton et al. (1999); Anon. (2002) Arabian oryx sanctuary, Oman, United Nations Environment Programme World Conservation Monitoring Centre; Anon. (2004) *Oryx leucoryx*. 4th International Conservation Workshop for the Threatened Fauna of Arabia, *2004 IUCN Red List of Threatened Species*

'There are hundreds of tortoises …': Merlen (1999)
No young Pinzón tortoises: Perry (1970)
Details of recovery of Pinzón eggs: MacFarland et al. (1974b)
Success of repatriation of Pinzón youngsters: Perry (1970); Dorst (1971)
'When the tortoise was found …': Snow (1963)
'The posterior slope of its carapace …': Anon. (1972) News from the
    Charles Darwin Research Station, Galápagos, *Noticias de Galápagos*
    19/20: 21–30
Details of Española captive-breeding effort: MacFarland et al. (1974b)
'The soil should be relatively fine …': MacFarland et al. (1974b)
Details of goat eradication on Pinta: Campbell et al. (2004)
'A meticulous search, aided by dogs …': Calvopiña (1985)
Discovery and translocation of 'No. 21': Thomas Fritts, personal
    communication; Fritts (1978)
'during the last century thousands of tortoises …': Fritts (1978)
Signs of tortoise breeding on Española: Marquez et al. (1991)
'Their appearance on Española …': Marquez et al. (1991)
Challenges of reintroducing golden lion tamarins: see Tudge (1992)
Repatriation of 1000th Española tortoise: Anon. (2000) 1,000 tortoises
    repatriated, *Galapagos Conservation Trust*, www.gct.org
Genetic contributions of Española founders: Milinkovitch et al. (2004)
Inbreeding experiments with fruit flies: see Spielman et al. (2004)
'Wildlife managers should strive to minimise inbreeding …': Spielman et
    al. (2004)
Rescue of the Mauritius kestrel: Groombridge et al. (2000)
Founder population of Isle Royale wolves: Wayne et al. (1991)
'It's more than likely that all the tortoise populations …': Linda Cayot,
    personal communication
'So it seems that on the last trip …': Evans (1990)
Judas goats: Campbell et al. (2004); Karl Campbell, personal
    communication
'From what we can gather it's malice': Karl Campbell, personal
    communication
Launch of the Isabela Project: Cayot (1998)
Proxy introduction of Aldabra tortoises to Mauritius and Curieuse: Justin
    Gerlach, personal communication
Rewilding North America: Donlan et al. (2005)
Proposal to introduce Española tortoises to Pinta: Campbell, personal
    communication
'If, in 20 years time, somebody decided …': Ole Hamann, personal
    communication

Chapter 10 – Faking organisms

'Given the minimal risk …': Fritts (2002)

Electroejaculation of wild elephants: Thomas Hildebrandt, personal
    communication
Electroejaculation of tortoises and turtles: Carrol Platz Jr, personal
    communication
'That was probably the biggest problem ...': Carrol Platz Jr, personal
    communication
Electroejaculation of Madagascar ploughshare tortoises: Juvik et al. (1991)
'There were good concentrations...': Carrol Platz Jr, personal
    communication
'Electroejaculation is not something they would ever do': Linda Cayot,
    personal communication
Cryopreservation of sperm, eggs and embryos: Pickard and Holt (2004);
    Bill Holt, personal communication
Black-footed ferret genome resource bank: Howard et al. (2003)
Artificial insemination in elephants: Hildebrandt, personal communication
'The best I ever got was 10% recovery...': Carrol Platz Jr, personal
    communication
Sperm storage in reptiles: Birkhead and Møller (1993); Fischer and Gist
    (1993); Palmer et al. (1998); Roques et al. (2004); Daniel Gist, personal
    communication
Obesity reduces sperm count: Jensen et al. (2004)
Nutritional study on Lonesome George: Oftedal (1994); Olav Oftedal,
    personal communication
Luteinizing hormone in elephants: Hildebrandt et al. (2003); Hildebrandt,
    personal communication
Reptile reproduction: Schramm et al. (1999); Lance (2003); Valentine
    Lance and Beatrix Schramm, personal communication
'I would not recommend it at all ...': Beatrix Schramm, personal
    communication

Chapter 11 – Clones and chimeras

'Cloning could be considered ...': Charles Darwin Research Station (2002)
Overview of cloning history: see Brownlee, C., (pnas.org/misc/classics4.
    shtml#Ref7)
Spemann's salamander: Spemann (1938)
Briggs and King's frogs: Briggs and King (1952)
Gurdon's frog clone: Gurdon (1962)
Cloning Dolly: Wilmut et al. (1997)
Cloning endangered species: Ryder (2002); Critser et al. (2003)
San Diego's Frozen Zoo: see cres.sandiegozoo.org/; Oliver Ryder, personal
    communication
'It has the potential of minimizing loss ...': Ryder and Benirschke (1997)
'flagship research' and 'exciting leading-edge science of international

significance': Anon. (1999) Cloning of extinct huia bird approved, Cable News Network, 20 July 1999

Cloning the thylacine: see www.amonline.net.au/thylacine/index.htm

'If we're lucky enough we'll find it': Larry Agenbroad, personal communication

'The condition of the skull ...': Agenbroad, personal communication

'the first complete sequence of the mitochondrial genome ...': Anon. (2005) The sequence of the complete mitochondrial DNA of the Yukagir Mammoth has been determined! EXPO 2005, 17 June

'It's full of holes ...': Crichton, M. and Koepp, D. (1993)

Degradation of ancient DNA: Alan Cooper, personal communication

Cloning the argali: White et al. (1999)

Cloning the gaur: Lanza et al. (2000)

'This study presents exciting opportunities ...': Lanza et al. (2000)

Cloning the mouflon: Loi et al. (2001); Pasqualino Loi, personal communication

Perry's chick: Perry (1988)

Creating avian chimeras: Petitte et al. (1997, 2004); James Petitte, personal communication; see also Graves (2001)

Fishy chimeras: Takeuchi et al. (2004)

Cloning mules: Don Jacklin and Gordon Woods, personal communication

'Once the cameras start clicking ...': Gordon Woods, personal communication

Brave New World: Huxley (1932)

'The prophecies I made in 1931 are coming true ...': Huxley (1958)

BIBLIOGRAPHY AND FURTHER READING

This bibliography contains all the citations made in Notes and Sources and several other useful texts consulted during the preparation of *Lonesome George*.

Adsersen, H. (1976) A botanist's notes on Pinta. *Noticias de Galápagos,* **24**, 26–7.

Adsersen, H. (1989) The rare plants of the Galápagos Islands and their conservation. *Biological Conservation,* **47**(1), 49–77.

Agassiz, A. (1892) Reports on the Dredging Operations off the West Coast of Central America to the Galápagos, to the West Coast of Mexico, and in the Gulf of California, in Charge of ALEXANDER AGASSIZ, carried on by the U.S. Fish Commission Steamer 'Albatross,' LIEUT. COMMANDER Z.L. TANNER, U.S.N., Commanding. *Bulletin of the Museum of Comparative Zoology Harvard,* **23**, 1–89.

Agenbroad, L. D. (2005) North American Proboscideans: mammoths: the state of knowledge, 2003. *Quaternary International,* **126–8**, 73–92.

Altamirano, M. et al. (2005) The application of the adaptive principle to the management and conservation of *Isostichopus fuscus* in the Galápagos Marine Reserve. *FAO Fisheries Technical Paper,* **463**, 247–58.

Arrowsmith, J. (1839) *South America, from Original Documents, including the Survey by the Officers of H. M. Ships* Adventure *and* Beagle. Dedicated to Captain R. Fitz Roy, R. N. (see www.galapagos.to).

Bagemihl, B. (2000) *Animal Exuberance: Animal Homosexuality and Natural Diversity.* Stonewall Inn Editions.

Balmford, A. and Whitten, T. (2003) Who should pay for tropical conservation, and how could the costs be met? *Oryx,* **37**(2), 238–50.

Barlow, N. (ed.) (1934) *Charles Darwin's Diary of the Voyage of H.M.S. 'Beagle'.* Cambridge University Press.

Barry, J. E. (1995) *Seizure of Galápagos National Park and Charles Darwin Research Station.* Charles Darwin Foundation.

Baur, G. (1889) The Gigantic Land Tortoises of the Galápagos Islands. *American Naturalist,* **23**(276), 1039–57.

Baur, G. (1890) On the classification of Testudinata. *American Naturalist,* **24**(282), 530–6.

Baur, G. (1891a) The Galápagos islands. Paper presented at the American Antiquarian Society, 21 October 1891.

Baur, G. (1891b) On the origin of the Galápagos Islands. *American Naturalist*, **25**(291), 217–29.

Baur, G. (1897) New observations on the origin of the Galápagos Islands, with remarks on the geological age of the Pacific Ocean. *American Naturalist*, **31**(368), 661–80.

Beck, R. H. (1902) In the home of the giant tortoise. *7th Annual Report of the New York Zoological Society*, **7**, 160–74.

Beebe, W. (1923) Galápagos reptiles and birds in the Zoological Park. *Zoological Society Bulletin*, **26**(5), 99–106.

Beebe, W. (1924) *Galápagos: World's End*. New York and London: G.P. Putnam's Sons.

Beheregaray, L. B. et al. (2003a) Genetic divergence, phylogeography and conservation units of giant tortoises from Santa Cruz and Pinzón, Galápagos Islands. *Conservation Genetics*, **4**(1), 31–46.

Beheregaray, L. B. et al. (2003b) Genes record a prehistoric volcano eruption in the Galápagos. *Science*, **302**(5642), 75.

Beheregaray, L. B. et al. (2004) Giant tortoises are not so slow: rapid diversification and biogeographic consensus in the Galápagos. *Proceedings of the National Academy of Sciences*, **101**(17), 6514–19.

Bensted-Smith, R. (1998) The Special Law for Galápagos. *Noticias de Galápagos*, **59**, 6.

Bensted-Smith, R. (ed.) (1999) *A Biodiversity Vision for the Galápagos Islands*. The Charles Darwin Foundation/World Wildlife Fund.

Berlanga, Bishop Frey Tomás de (1535) Letter to His Majesty ... describing his Voyage from Panamá to Puerto Viejo. In *Coleccion de Documentos Ineditos relativos al Descubrimiento, Conquista y Organizacion de las Antiguas Posesiones Españolas de América y Oceania*, Tomo XLI, Cuaderno II. Madrid. Imprenta de Manuel G. Hernandez (1884, pp. 538–44).

Birkhead, T. R. and Møller, A. P. (1993) Sexual selection and sperm storage. *Biological Journal of the Linnean Society*, **50**, 295–311.

Boersma, D. et al. (2005) Living laboratory in peril. *Science*, **308**(5724), 925.

Briggs, R. and King, T. J. (1952) Transplantation of living nuclei from blastula cells into enucleated frogs eggs. *Proceedings of the National Academy of Sciences*, **38**, 455–63.

Buddenhagen, C. E. et al. (2004) The control of a highly invasive tree *Cinchona pubescens* in Galápagos. *Weed Technology*, **18**, 1194–202.

Bunce, M. et al. (2005) Ancient DNA provides new insights into the evolutionary istory of New Zealand's extinct giant eagle. *Public Library of Science Biology,* **3**(1), e9.

Burns, C. E. et al. (2003) The origin of captive Galápagos tortoises based on DNA analysis: implications for the management of natural populations. *Animal Conservation,* **6**, 329–37.

Caccone, A. et al. (1999) Origin and evolutionary relationships of giant Galápagos tortoises. *Proceedings of the National Academy of Sciences,* **96**(23), 13223–8.

Caccone, A. et al. (2002) Phylogeography and history of giant Galápagos tortoises. *Evolution,* **56**(10), 2052–66.

Caccone, A. et al. (2004) Extreme difference in rate of mitochondrial and nuclear DNA evolution in a large ectotherm, Galápagos tortoises. *Molecular Phylogenetics and Evolution,* **31**(2), 794–8.

Calvopiña, M. (1985) *Annual Report 1984–1985.* Quito, Ecuador: Department of Introduced Mammals, Charles Darwin Research Station.

Campbell, K. et al. (2004) Eradication of feral goats *Capra hircus* from Pinta Island, Galápagos, Ecuador. *Oryx,* **38**(3), 328–33.

Campbell, K. H. S. et al. (1994) Improved development to blastocyst of ovine nuclear transfer embryos reconstructed during the presumptive S-phase of enucleated activated oocytes. *Biology of Reproduction,* **50**, 1385–93.

Campbell, K. H. S. et al. (1996) Sheep cloned by nuclear transfer from a cultured cell line. *Nature,* **380**, 64–6.

Caro, T. M. et al. (2004) Preliminary assessment of the flagship species concept at a small scale. *Animal Conservation,* **7**, 63–70.

Causton, C. E. (2005) Ladybugs to the rescue. *Galápagos News,* **21**, 1–2.

Causton, C. E. et al. (2004) Feeding range studies of *Rodolia cardinalis* (Mulsant), a candidate biological control agent of *Icerya purchasi* Maskell in the Galápagos islands. *Biological Control,* **29**(3), 315–25.

Cayot, L. (1998) The Isabela project: off and running. *Noticias de Galápagos,* **59**.

Cayot, L. J. (1991) The passing of two beloved reptiles: Onan and Chiquita. *Noticias de Galápagos,* **50**, 5–7.

Cayot, L. J. and Lewis, E. (1994) Recent increase in killing of giant tortoises on Isabela Island. *Noticias de Galápagos,* **54**, 2–7.

Chambers, P. (2004a) *A Sheltered Life: The Unexpected History of the Giant Tortoise.* London: John Murray.

Chambers, P. (2004b) The origin of Harriet. *New Scientist,* 11 September.

Charles Darwin Research Station (2002) Lonesome George: Last of a Species?

Chen, J. (2003) Overview of sea cucumber farming and sea ranching practices in China. *SPC Beche-de-Mer Information Bulletin,* **18**, 18–23.

Cherfas, J. (1995) Goats must go to save the Galápagos tortoises. *New Scientist,* 15 April.

Christie, D. M. et al. (1992) Drowned islands downstream from the Galápagos hotspot imply extended speciation times. *Nature,* **355**(6357), 246–8.

Ciofi, C. et al. (2002) Microsatellite analysis of genetic divergence among populations of giant Galápagos tortoises. *Molecular Ecology,* **11**, 2265–83.

Colnett, J. (1798) *A Voyage to the South Atlantic and Round Cape Horn into the Pacific Ocean, for the Purpose of Extending the Spermaceti Whale Fisheries, and other objects of commerce, by ascertaining the ports, bays, harbours, and anchoring births, in certain islands and coasts in those seas at which the ships of the British merchants might be refitted.* London: Bennett, W.

Cookson, W. E. (1876a) Extract from a Report by Commander Cookson, R.N., of a visit by H.M.S. 'Peterel' to the Galápagos Islands in July 1875, which had been communicated to him by the First Lord of the Admiralty. Read out at the *Proceedings of the Zoological Society of London,* 1 February 1876.

Cookson, W. E. (1876b) Letter from Commander W. E. Cookson, R.N. [to Alfred Günther May 29 1876]. Read out at the *Proceedings of the Zoological Society of London,* 20 June 1876.

Cope, E. D. (1890) Scientific results of explorations by the U.S. fish commission steamer Albatross. No. III. – Report on the batrachians and reptiles collected in 1887–'88. *Proceedings of the US National Museum,* **12**, 141–7.

Corely Smith, G. T. (1976) Saving the giant tortoises of the Galápagos from extinction. *Noticias de Galápagos,* **25**, 13–19.

Corely Smith, G. T. (1981) The significance of the Perry Isthmus. *Noticias de Galápagos,* **33**, 24.

Corley Smith, H. E. (1990) Roger Perry. Research Station Director 1964–1970. *Noticias de Galápagos,* **49**, 14–16.

Cowley, W. A. (1687–89?) *Cowley's Voyage Round the World.* London: British Library Sloane MS. 1050.

Crichton, M. and Koepp, D. (1993) Screenplay for *Jurassic Park.* Universal Pictures and Amblin Entertainment.

Critser, J. K. et al. (2003) Application of nuclear transfer technology to wildlife species. In Holt, W. V. et al. (eds) *Reproductive Science and Integrated Conservation* (pp. 195–208). Cambridge: Cambridge University Press.

Cruz, E. (1997a) Park warden wounded by bullet in confrontation between illegal sea cucumber fishermen and patrol personnel of the Galápagos National Park. *Noticias de Galápagos*, **58**, 2.

Cruz, E. (1997b) Peaceful demonstration to reject violence in Galápagos. *Noticias de Galápagos*, **58**, 2–3.

Cruz, F. et al. (2005) Conservation action in the Galápagos: feral pig (*Sus scrofa*) eradication from Santiago Island. *Biological Conservation*, **121**(3), 473–8.

Cruz, M. (1994) The story of the discovery of the tortoise 'Lonesome George' on Pinta Island. *Noticias de Galápagos*, **53**, 15–18.

Culver, M. et al. (2000) Genomic ancestry of the American puma (*Puma concolor*). *Journal of Heredity*, **91**, 86–197.

Darwin, C. (1839) *Journal of Researches into the Geology and Natural History of the various Countries Visited by H.M.S. Beagle* (1st edn). London: Henry Colburn.

Darwin, C. (1845) *Journal of Researches into the Natural History and Geology of the Countries Visited During the Voyage Round the World by H.M.S. Beagle* (2nd edn). London: Henry Colburn.

Darwin, C. (1859) *On The Origin of Species by Means of Natural Selection*. London: John Murray.

Darwin (1890) *Journal of Researches into the Natural History and Geology of the countries visited during the voyage round the world of H.M.S. Beagle.* (11th edn) London: John Murray.

Darwin (1899?) *Journal of Researches into the Natural History and Geology of the countries visited during the voyage of H.M.S. "Beagle" round the world.* London: Ward, Lock & Co., Limited.

Daugherty, C. H. et al. (1990) Neglected taxonomy and continuing extinctions of tuatara (*Sphenodon*). *Nature*, **347**(6289), 177–9.

de Groot, R. S. (1983) Tourism and conservation in the Galápagos Islands. *Biological Conservation*, **26**(4), 291–300.

de Queiroz, A. (2005) The resurrection of oceanic dispersal in historical biogeography. *Trends in Ecology and Evolution*, **20**(2), 68–73.

Desender, K. et al. (2002) *Calleida migratoria* Casale, new species (Coleoptera : Carabidae), a newly introduced ground beetle in the Galápagos Islands, Ecuador. *Coleopterists Bulletin*, **56**(1), 71–8.

Donlan, C. J. et al. (2005) Re-wilding North America. *Nature*, **436**, 913–14.

D'Orso, M. (2003) *Plundering Paradise: The Hand of Man on the Galápagos Islands*. New York: Perennial.

Dorst, J. (1964) Découverte d'une population de tortues a San Cristóbal. *Noticias de Galápagos*, **4**, 19.

Dorst, J. (1971) Elevage de tortues en captivité. *Noticias de Galápagos*, **17**, 18.

Dowler, R. C. et al. (2000) Rediscovery of rodents (Genus *Nesoryzomys*) considered extinct in the Galápagos Islands. *Oryx*, **34**(2), 109.

Edgar, G. J. et al. (2004) Bias in evaluating the effects of marine protected areas: the importance of baseline data for the Galápagos Marine Reserve. *Environmental Conservation*, **31**(3), 212–18.

Estes, E. et al. (2000) Darwin in Galápagos: his footsteps through the archipelago. *Notes and Records of the Royal Society of London*, **54**(3), 343–68.

Evans, D. (1990) *Informe del Director*. Quito, Ecuador: Reunión del Consejo Ejecutivo de la Fundación Charles Darwin.

Evans, M. J. et al. (1999) Mitochondrial DNA genotypes in nuclear transfer-derived cloned sheep. *Nature Genetics*, **23**, 90–3.

Farr, T. and Kobrick, M. (2001) The Shuttle radar topography mission. *Eos Trans. American Geophys. Union* 82, 47.

Farragut, L. (1879) *The Life of David Glasgow Farragut, First Admirial of the United States Navy*. New York: D. Appleton and Company.

Ferber, D. (2000) Galápagos station survives latest attack by fishers. *Science*, **290**(5499), 2059–61.

Ferraro, P. J. and Kiss, A. (2002) Direct payments to conserve biodiversity. *Science*, **298**(5599), 1718–19.

Ferraro, P. J. and Simpson, R. D. (2003) Protecting forests and biodiversity: Are investments in eco-friendly production activities the best way to protect endangered ecosystems and enhance rural livelihoods? Paper presented at the International Conference on Rural Livelihoods, Forests and Biodiversity, Bonn, Germany, 19–23 May 2003.

Fessl, B. and Tebbich, S. (2002) *Philornis downsi* – a recently discovered parasite on the Galápagos archipelago – a threat for Darwin's finches? *Ibis*, **144**, 445–51.

Fischer, E. and Gist, D. H. (1993) Fine structure of sperm containing glands of the box turtle oviduct. *Journal of Reproduction and Fertility*, **97**, 463–8.

Fitzpatrick, J. W. et al. (2005) Ivory-billed woodpecker (*Campephilus principalis*) persists in continental North America. *Science*, **308**(5727), 1460–2.

Fitzroy, R. (1839) *Narrative of the Surveying Voyages of his Majesty's Ships*

Adventure and Beagle, between the Years 1826 and 1836, describing their Examination of the Southern Shores of South America, and the Beagle's Circumnavigation of the Globe. London: Henry Colburn.

Frankham, R. (2003) Genetics and conservation biology. *Biologies*, **326**, S22–S29.

Frankham, R. et al. (2002) *Introduction to Conservation Genetics.* Cambridge: Cambridge University Press.

Freeberg, T. M. (2000) Culture and courtship in vertebrates: a review of social learning and transmission of courtship systems and mating patterns. *Behavioural Processes*, **51**(1–3), 177–92.

Freedberg, S. and Wade, M. J. (2001) Cultural inheritance as a mechanism for population sex-ratio bias in reptiles. *Evolution*, **55**(5), 1049–55.

Freeman, R. B. (1985) Darwin in the Galápagos. *Noticias de Galápagos*, **42**, 15–16.

Fritts, T. H. (1978) Española tortoise returns to Galápagos. *Noticias de Galápagos*, **28**, 17–18.

Fritts, T. H. (1983) Morphometrics of Galápagos tortoises: evolutionary implications. In Bowman, R. I. et al. (eds) *Patterns of Evolution in Galápagos Organisms* (pp. 107–22). San Francisco, CA: Pacific Division, American Association for the Advancement of Science.

Fritts, T. H. (1984) Evolutionary divergence of giant tortoises in Galápagos. *Biological Journal of the Linnean Society*, **21**, 165–76.

Fritts, T. H. (2002) The Pinta Island tortoise population: background, present context, and opportunities for recovery of this most endangered giant tortoise population. Puerto Ayora: Charles Darwin Research Station.

Fritts, T. H. and Fritts, P. H. (eds) (1982) *Race with Extinction: Herpetological Notes of J. R. Slevin's Journey to the Galápagos 1905–1906.* Lawrence, Kansas: Herpetologists' League.

Fritts, T. H. et al. (1998) Progress and priorities in research for the conservation of reptiles. Paper presented at Science for Conservation in Galápagos, Belgium.

Gaidos, S. (2005) Keeping it in the family. *New Scientist*, 13 August.

Gerlach, J. (1998) *Famous Tortoises.* Cambridge: J. Gerlach.

Ginsberg, J. R. (2001) Flagship panda. *Trends in Ecology and Evolution*, **16**(1), 56–7.

Gould, C. G. (2004) *The Remarkable Life of William Beebe: Explorer and Naturalist.* Washington, D.C.: Island Press/Shearwater Books.

Graves, A. (2001) Clone farm. *New Scientist*, 18 August.

Grigioni, S. (1993) *Georges le solitaire: la tortue géante orpheline.* Puerto Ayora: Charles Darwin Research Station.

Groombridge, J. et al. (2000) 'Ghost' alleles of the Mauritius kestrel. *Nature*, **403**(6770), 616.

Günther, A. (1875) Description of the living and extinct races of gigantic land-tortoises. Part 1–2. *Philosophical Transactions of the Royal Society of London, Biological Sciences*, **165**, 251–84.

Günther, A. (1877a) Account of the Zoological Collection made during the visit of H.M.S. Peterel to the Galápagos Islands. Communicated by Dr. Albert Günther, FRS, VPZS, Keeper of the Zoological Department, British Museum. *Proceedings of the Zoological Society of London*, 64–92.

Günther, A. (1877b) *The Gigantic Land-Tortoises (Living and Extinct) in the Collection of the British Museum.* London: Taylor and Francis.

Günther, A. (1896) *Testudo ephippium. Novitates Zoologicae*, **3**(4), 329–34.

Gurdon, J. B. (1962) The developmental capacity of nuclei taken from intestinal epithelium cells of feeding tadpoles. *Journal of Embryology and Experimental Morphology*, **10**, 622–40.

Gurdon, J. B. and Colman, A. (1999) The future of cloning. *Nature*, **402**(6763), 743–6.

Hall, B. (1826) *Extracts from a Journal, Written on the Coasts of Chili, Peru, and Mexico, in the years 1820, 1821, 1822.* Edinburgh: Archibald Constable and Co.

Hamann, O. (1975) Vegetational changes in the Galápagos Islands during the period 1966–1973. *Biological Conservation*, **7**(1), 37–59.

Hamann, O. (1978) Recovery of vegetation on Pinta and Santa Fe Islands. *Noticias de Galápagos*, **27**, 19–20.

Hamann, O. (1979) Regeneration of vegetation on Santa Fe and Pinta islands, Galápagos, after the eradication of goats. *Biological Conservation*, **15**(3), 215–35.

Hamann, O. (1993) On vegetation recovery, goats and giant tortoises on Pinta Island, Galápagos, Ecuador. *Biodiversity and Conservation*, **2**(2), 138–51.

Hamann, O. (2001) Demographic studies of three indigenous stand-forming plant taxa (*Scalesia*, *Opuntia* and *Bursera*) in the Galápagos Islands, Ecuador. *Biodiversity and Conservation*, **10**(2), 223–50.

Hardin, G. (1968) The tragedy of the commons. *Science*, **162**, 1243–8.

Harris, J. (1744) *Navigantium atque Itinerantium Bibliotheca.* (2nd and 3rd edns) Vol. 1, p. 79 (see www.galapagos.to).

Hearn, A. et al. (2005) Population dynamics of the exploited sea cucumber *Isostichopus fuscus* in the western Galápagos Islands, Ecuador. *Fisheries Oceanography,* **14**(5), 377–85.

Henderson, S. J. and Whittaker, R. J. (2002) Islands. *Encylopedia of Life Sciences.*

Heylings, P. (1999) *Summary of the Changes and Advances in the Management and Protection of the Galápagos Marine Reserve.* Puerto Ayora: Galápagos National Park/Charles Darwin Research Station.

Hickman, J. (1991) *The Enchanted Islands: The Galápagos Rediscovered.* Oswestry, Shropshire: Anthony Nelson.

Hildebrandt, T. B. et al. (2003) Ultrasound for analysis of reproductive function in wildlife species. In Holt, W. V. et al. (eds) *Reproductive Science and Integrated Conservation* (pp. 166–82). Cambridge: Cambridge University Press.

Hooker, J. D. (1851) On the vegetation of the Galápagos archipelago. *Transactions of the Linnean Society of London,* **20**, 235–62.

Howard, J. et al. (2003) Black-footed ferret: model for assisted reproductive technologies contributing to *in situ* conservation. In Holt, W. V. et al. (eds) *Reproductive Science and Integrated Conservation* (pp. 249–66). Cambridge: Cambridge University Press.

Huxley, A. (1932) *Brave New World.* London: Flamingo.

Huxley, A. (1958) *Brave New World Revisited.* London: Flamingo.

Ingold, D. J. (1989) Look, Mom, no cavities. *The Living Bird Quarterly* (summer), 24–8.

IUCN/SSC (1988) *Guidelines for Reintroductions.* Gland, Switzerland and Cambridge: IUCN.

Jackson, P. F. R. (1974) New lease of life for Darwin's giant tortoises, Galápagos Islands. *Biological Conservation,* **6**(1), 63–4.

Jensen, T. K. et al. (2004) Body mass index in relation to semen quality and reproductive hormones among 1,558 Danish men. *Fertility and Sterility,* **82**(4), 863–70.

Jones, P. (1995) Galápagos marine nature-reserves threatened. *Marine Pollution Bulletin,* **30**(11), 684–5.

Juniper, T. (2002) *Spix's Macaw: the Race to Save the World's Rarest Bird.* London and New York: Fourth Estate.

Juvik, J. O. et al. (1991) Captive husbandry and conservation of the Madagascar ploughshare tortoise, *Geochelone yniphora.* Paper presented at the First International Symposium on Turtles and Tortoises: Conservation and Captive Husbandry.

Kaiser, J. (2001) Galápagos takes aim at alien invaders. *Science,* **293**(5530), 590–2.

Keogh, J. S. et al. (2005) Rapid and related origin of insular gigantism and dwarfism in Australian tiger snakes. *Evolution*, **59**(1), 226–33.

Kerr, S. et al. (2004) Migration and the environment in the Galápagos: an analysis of economic and policy incentives driving migration, potential impacts from migration control, and potential policies to reduce migration pressure (No. Motu Working Paper 03–17): Motu Economic and Public Policy Research.

Kerr, S. A. (2005) What is small island sustainable development about? *Ocean and Coastal Management*, **48**(7–8), 503–24.

Keynes, R. D. (ed.) (2000) *Charles Darwin's Zoology Notes and Specimen Lists from H.M.S. Beagle*. Cambridge University Press.

Kizirian, D. et al. (2004) Evolution of Galápagos Island lava lizards (Iguania: Tropiduridae: Microlophus). *Molecular Phylogenetics and Evolution*, **32**(3), 761–9.

Kruger, P. (2005) The role of ecotourism in conservation: panacea or Pandora's box? *Biodiversity and Conservation*, **14**(3), 579–600.

Lance, V. A. (2003) Reptile reproduction and endocrinology. In Holt, W. V. et al. (eds) *Reproductive Science and Integrated Conservation* (pp. 339–58). Cambridge: Cambridge University Press.

Lanza, R. P. et al. (2000) Cloning of an endangered species (*Bos gaurus*) using interspecies nuclear transfer. *Cloning*, **2**(2), 79–90.

Larson, E. J. (2001) *Evolution's Workshop: God and Science on the Galápagos Islands*. London: Penguin.

Lazell, J. (2002) Restoring vertebrate animals in the British Virgin islands. *Ecological Restoration*, **20**(3), 179–85.

Leader-Williams, N. (2003) Regulation and protection: successes and failures in rhinoceros conservation. In Oldfield, S. (ed.) *The Trade in Wildlife: Regulation for Conservation* (pp. 89–99). Earthscan Publications.

Leader-Williams, N. and Dublin, H. T. (2000) Charismatic megafauna as 'flagship species'. In Entwistle, A. and Dunstone, N. (eds) *Priorities for the Conservation of Mammalian Diversity: Has the Panda had its Day?* (pp. 53–81). Cambridge: Cambridge University Press.

Lemonick, M. D. and Dorfman, A. (1995) Can the Galápagos survive? *Time*, **146**(18), 80–2.

Ley, D. (2003) An assessment of energy and water in the Galápagos. Masters thesis, University of Colorado.

Lezama, V. et al. (2001) Sexual behavior and semen characteristics of rams exposed to their own semen or semen from a different ram on the vulva of the ewe. *Applied Animal Behaviour Science*, **75**(1), 55–60.

Line, L. (1994) Have wings, can't fly – ostriches and other non-flying birds. *International Wildlife*, 1 November.

Loi, P. et al. (2001) Genetic rescue of an endangered mammal by cross-species nuclear transfer using post-mortem somatic cells. *Nature Biotechnology*, 19(10), 962–4.

Loope, L. L. et al. (1988) Comparative conservation biology of Oceanic archipelagos. *Bioscience*, 38(4), 272–82.

Lyell, C. (1830) *Principles of Geology*. London: John Murray.

MacFarland, C. G. (1998) An analysis of nature tourism in the Galápagos Islands. Paper presented at Science for Conservation in the Galápagos, Belgium.

MacFarland, C. G. and Cifuentes, M. (1996) Biodiversity conservation and human population impacts in the Galápagos Islands, Ecuador. In Dompka, V. (ed.) *Human Population, Biodiversity and Protected Areas: Science and Policy Issues. Report of a Workshop April 20–21, 1995, Washington, D.C.* Washington, DC: American Association for the Advancement of Science.

MacFarland, C. G. et al. (1974a) The Galápagos giant tortoises (*Geochelone elephantopus*) Part I: status of the surviving populations. *Biological Conservation*, 6(2), 118–33.

MacFarland, C. G. et al. (1974b) The Galápagos giant tortoises (*Geochelone elephantopus*) Part II: conservation methods. *Biological Conservation*, 6(3), 198–212.

Maddox, J. (1993) Bringing the extinct dodo back to life. *Nature*, 365, 291.

Magsalay, P. et al. (1995) Extinction and conservation on Cebu. *Nature*, 373, 294.

Malavasi, E. O. and Kellenberg, J. (2002) Program of Payments for Ecological Services in Costa Rica. Paper presented at the International Expert Meeting on Forest Landscape Restoration. Heredia, Costa Rica, 27–28 February.

Malik, S. et al. (1997) Pinniped penises in trade: a molecular-genetic investigation. *Conservation Biology*, 11(6), 1365–74.

Marquez, C. (1986) The giant tortoises and the great fire on Isabela. *Noticias de Galápagos*, 44, 8.

Marquez, C. et al. (1987) The giant tortoise conservation program. *Noticias de Galápagos*, 45, 17–18.

Marquez, C. et al. (1991) A 25-year management program pays off: repatriated tortoises on Española reproduce. *Noticias de Galápagos*, 50, 17–18.

Marquez, C. et al. (1994) The fire of 1994 and herpetofauna of Southern Isabela. *Noticias de Galápagos*, **54**, 8–10.

Marquez, C. et al. (2001) *The Captive Rearing of Galápagos Tortoises: an Operative Manual*. Puerto Ayora: Charles Darwin Research Station.

Martin, E. and Stiles, D. (2005) *Ivory Markets in Europe*. Care for the Wild International and Save the Elephants.

Melville, H. (1854) The *Encantadas* or, Enchanted Isles. *Putnam's Monthly Magazine*, **3**.

Merlen, G. (1993) Pepino war, 1992. *Noticias de Galápagos*, **52**, 3.

Merlen, G. (1999) *Restoring the Tortoise Dynasty*. Quito, Ecuador: Charles Darwin Foundation.

Metzger, S. and Marlow, R. W. (1986) The status of the Pinzón Island giant tortoise. *Noticias de Galápagos*, **43**, 18–20.

Milinkovitch, M. C. et al. (2004) Genetic analysis of a successful repatriation programme: giant Galápagos tortoises. *Proceedings of the Royal Society – Series B*, **271**, 341–5.

Moloney, S. and Vanderwoude, C. (2002) Red imported fire ants. A threat to eastern Australia's wildlife? *Ecological Management and Restoration*, **3**(3), 167–75.

Narain, S. et al. (2005) *Joining the Dots*. New Delhi: Tiger Task Force, Ministry of the Environment and Forests.

Nowak, R. (1994) Fire threatens Galápagos tortoises. *Science*, **264**(5159), 651.

Oftedal, O. T. (1994) *Galápagos Tortoises: Nutrition and Suggested Diet Changes*. Puerto Ayora: Charles Darwin Research Station.

Okey, T. A. et al. (2004) A trophic model of a Galápagos subtidal rocky reef for evaluating fisheries and conservation strategies. *Ecological Modelling*, **172**(2–4), 383–401.

Ophir, A. G. and Galef, B. G. (2003) Female Japanese quail affiliate with live males that they have seen mate on video. *Animal Behaviour*, **66**, 369–75.

Osborn, F. V. and Parker, G. E. (2002) *Living with Elephants II: A Manual*. Harare: Elephant Pepper Developments.

Osborn, F. V. and Parker, G. E. (2003) Towards an integrated approach for reducing the conflict between elephants and people: a review of current research. *Oryx*, **37**(1), 1–5.

Pain, B. et al. (1999) Chicken embryonic stem cells and transgenic strategies. *Cells Tissues Organs*, **165**(3–4), 212–19.

Pain, B. et al. (1997) Avian embryonic stem cells: a new approach for efficient avian transgenesis. *European Journal of Cell Biology*, **74**, 121–121.

Palmer, K. S. et al. (1998) Long-term sperm storage in the desert tortoise. *Copeia*, **3**, 702–5.

Parsons, K. M. et al. (1999) Amplifying dolphin mitochondrial DNA from faecal plumes. *Molecular Ecology*, **8**, 1753–68.

Parsons, K. M. et al. (2003) Kinship as a basis for alliance formation between male bottlenose dolphins, *Tursiops truncatus*, in the Bahamas. *Animal Behaviour*, **66**(1), 185–94.

Pearce, F. (1995) Galápagos tortoises under siege. *New Scientist*, 16 September.

Perry, M. (1988) A complete culture system for the chick embryo. *Nature*, **331**, 70–2.

Perry, R. (1964) Santa Cruz tortoise reserve. *Noticias de Galápagos*, **4**, 19–20.

Perry, R. (1970) Tortoise rearing in the Galápagos Islands. *Noticias de Galápagos*, **15/16**, 3–7.

Petitte, J. N. et al. (1997) The origin of the avian germ line and transgenesis in birds. *Poultry Science*, **76**(8), 1084–92.

Petitte, J. N. et al. (2004) Avian pluripotent stem cells. *Mechanisms of Development*, **121**(9), 1159–68.

Pickard, A. R. and Holt, W. V. (2004) Cryopreservation as a supporting measure in species conservation; 'not the frozen zoo!' In Benson, E. et al. (eds) *Life in the Frozen State* (pp. 393–413). Baton Rouge: CRC Press.

Pieau, C. and Dorizzi, M. (2004) Oestrogens and temperature-dependent sex determination in reptiles: all is in the gonads. *Journal of Endocrinology*, **181**, 367–77.

Pires, A. E. and Fernandes, M. L. (2003) Last lynxes in Portugal? Molecular approaches in a pre-extinction scenario. *Conservation Genetics*, **4**, 525–32.

Porter, D. (1813) Letter. In Brannan, J. (ed.) *Official Letters of the Military and Naval Officers of the United States During the War with Great Britain in the Years 1812, 13, 14, and 15 With Some Additional Letters and Documents Elucidating the History of that Period* (pp. 175–6). Washington, 1823.

Porter, D. (1815) *Journal of a Cruise made to the Pacific Ocean, by Captain David Porter, in the United States Frigate ESSEX, in the years 1812, 1813, and 1814. Containing Descriptions of the Cape de Verd Islands, Coasts of Brazil, Patagonia, Chili, and Peru, and of the Gallapagos Islands; also, A full Account of the Washington Groupe* [sic] *of Islands, the Manners, Customs, and Dress of the Inhabitants, etc., etc.* Philadelphia: Bradford and Inskeep.

Powell, J. R. and Gibbs, J. P. (1995) A report from Galápagos. *Trends in Ecology and Evolution*, **10**(9), 351–4.

Pritchard, P. C. H. (1977) Three, two, one tortoise. *Natural History Magazine*, **86**, 91–100.

Pritchard, P. C. H. (1979) *Encyclopedia of Turtles*. Jersey City: TFH Publications.

Pritchard, P. C. H. (1984) Further thoughts on 'Lonesome George'. *Noticias de Galápagos*, **39**, 20–3.

Pritchard, P. C. H. (1996) The Galápagos tortoises: nomenclatural and survival status. *Chelonian Research Monographs*, **1**.

Pritchard, P. C. H. (2004) *The 2003 Pinta Island Expedition*. Chelonian Research Institute/Galápagos National Park Service.

Ptak, G. et al. (2002) Preservation of the wild European mouflon: the first example of genetic management using a complete program of reproductive biotechnologies. *Biology of Reproduction*, **66**, 796–801.

Quammen, D. (1996) *The Song of the Dodo. Island Biogeography in an Age of Extinctions*. New York: Simon & Schuster.

Rassmann, K. (1997) Evolutionary age of the Galápagos iguanas predates the age of the present Galápagos Islands. *Molecular Phylogenetics and Evolution*, **7**(2), 158–72.

Reynolds, R. P. and Marlow, R. W. (1983) Lonesome George, the Pinta Island tortoise: a case of limited alternatives. *Noticias de Galápagos*, **37**, 14–17.

Romero, L. M. and Wikelski, M. (2002) Exposure to tourism reduces stress-induced corticosterone levels in Galápagos marine iguanas. *Biological Conservation*, **108**, 371–4.

Roque-Albelo, L. and Causton, C. E. (1999) El Niño and introduced insects in the Galápagos Islands: different dispersal strategies, similar effects. *Noticias de Galápagos*, **60**, 30–6.

Roques, S. et al. (2004) Microsatellite markers reveal multiple paternity and sperm storage in the Mediterranean spurthighed tortoise, *Testudo graeca*. *Canadian Journal of Zoology*, **82**, 153–9.

Rothschild, M. (1983) *Dear Lord Rothschild: Birds, Butterflies and History*. Philadelphia: Balaban Publishers.

Rothschild, W. (1915) The giant land tortoises of the Galápagos Islands in the Tring Museum. *Novitates Zoologicae*, **22**, 403–17.

Rothschild, W. and Hartert, E. (1899) A review of the ornithology of the Galápagos Islands. With notes on the Webster–Harris expedition. *Novitates Zoologicae*, **6**, 85–205.

Russello, M. A. et al. (2005) A cryptic taxon of Galápagos tortoise in conservation peril. *Biology Letters*, **1**, 287–90.

Ryder, O. A. (2002) Cloning advances and challenges for conservation. *Trends in Biotechnology*, **20**(6), 231–2.

Ryder, O. A. and Benirschke, K. (1997) The potential use of 'cloning' in the conservation effort. *Zoo Biology*, **16**, 295–300.

Schramm, B. G. et al. (1999) Steroid levels and reproductive cycle of the Galápagos tortoise, *Geochelone nigra*, living under seminatural conditions on Santa Cruz Island (Galápagos). *General and Comparative Endocrinology*, **114**(1), 108–20.

Shaw, G. K. (2000) Peter Pritchard: hero of the planet. *Orlando Sentinel*, 20 February.

Shepherd, S. A. et al. (2004) The Galápagos sea cucumber fishery: management improves as stocks decline. *Environmental Conservation*, **31**(2), 102–10.

Shine, R. (1999) Why is sex determined by nest temperature in many reptiles? *Trends in Ecology and Evolution*, **14**(5), 186–9.

Sinton, C. W. et al. (1996) Geochronology of Galápagos seamounts. *Journal of Geophysical Research – Solid Earth*, **101**(B6), 13689–700.

Slevin, J. R. (1959) The Galápagos Islands: a history of their exploration. *Occasional Papers of the California Academy of Sciences*, **25**, 1–150.

Smith, W. H. F. and Sandwell, D. T. (1997) Global seafloor topography from satellite altimetry and ship depth soundings. *Science* **277**, 1957–62.

Snell, H. et al. (2002) Current status of and threats to the terrestrial biodiversity of Galápagos. In Bensted-Smith, R. (ed.) *A Biodiversity Vision for the Galápagos Islands* (pp. 30–47). Puerto Ayora: Charles Darwin Foundation/World Wildlife Fund.

Snell, H. L. and Rea, S. (1999) The 1997–98 El Niño in Galápagos: can 34 years of data estimate 120 years of pattern? *Noticias de Galápagos*, **60**, 11–20.

Snow, D. (1963) The tortoise marking scheme, and other tortoise studies. *Noticias de Galápagos*, **2**, 9–10.

Snow, D. (1964) State of the other tortoise populations. *Noticias de Galápagos*, **3**, 19.

Sohlman, E. (2004) A bid to save the Galápagos of the Indian Ocean. *Science*, **303**(5665), 1753.

Spalton, J. A. et al. (1999) Arabian oryx reintroduction in Oman: successes and setbacks. *Oryx*, **33**(2), 168–75.

Spemann, H. (1938) *Embryonic Development and Induction*. Newhaven, CT: Yale University Press.

Spielman, D. et al. (2004) Does inbreeding and loss of genetic diversity decrease disease resistance? *Conservation Genetics*, 5(4), 439–48.

Steadman, D. W. and Martin, P. S. (2003) The late Quaternary extinction and future resurrection of birds on Pacific islands. *Earth-Science Reviews*, 61(1–2), 133–47.

Steadman, D. W. et al. (1991) Chronology of Holocene vertebrate extinction in the Galápagos Islands. *Quaternary Research*, 36(1), 126–33.

Stone, R. (1995) Fishermen threaten Galápagos. *Science*, 267(5198), 611–12.

Stronza, A. L. (2000) *Because it is Ours: Community-based Ecotourism in the Peruvian Amazon*. PhD thesis, University of Florida, Gainsville.

Sulloway, F. (1982) Darwin's conversion: the Beagle voyage and its aftermath. *Journal of the History of Biology*, 15, 325–96.

Sulloway, F. (1984) Darwin and the Galápagos. *Biological Journal of the Linnean Society*, 21, 29–59.

Sulloway, F. (1985) Darwin's 'dogged' genius: his Galápagos visit in retrospect. *Noticias de Galápagos*, 42, 7–14.

Summers, A. P. et al. (1999) Agassiz, Garman, Albatross, and the collection of deep-sea fishes. *Marine Fisheries Review*, 61(4), 58–68.

Takeuchi, Y. et al. (2004) Surrogate broodstock produces salmonids. *Nature*, 430, 629–30.

Thomas, L. (1995) *The Lives of a Cell: Notes of a Biology Watcher*. London: Penguin.

Thornton, I. (1971) *Darwin's Islands: A Natural History of the Galápagos*. New York: Natural History Press.

Toral-Granda, M. V. and Martinez, P. C. (2005) Population density and fishery impacts on the sea cucumber (*Isostichopus fuscus*) in the Galápagos marine reserve. *FAO Fisheries Technical Paper*, 463, 91–100.

Toral-Granda, V. (2005) Requiem for the Galápagos sea cucumber fishery? *Beche-de-Mer Information Bulletin*, 21, 5–8.

Tortoise and Freshwater Turtle Specialist Group (1996) *Geochelone nigra* ssp. *abingdoni*. In *2004 IUCN Red List of Threatened Species*.

Townsend, C. H. (1924a) Impending extinction of the Galápagos tortoises. *Bulletin of the New York Zoological Society*, 27(2), 55–6.

Townsend, C. H. (1924b) New information on the Galápagos tortoises. *Bulletin of the New York Zoological Society*, 27(4), 89.

Townsend, C. H. (1925a) The Galápagos tortoises in their relation to the whaling industry. A study of old logbooks. *Zoologica*, 4(3), 55–135.

Townsend, C. H. (1925b) The whaler and the tortoise. *The Scientific Monthly*, 166–72.

Townsend, C. H. (1928) Propagation of the giant tortoise in the United States. *Science*, **68**(1750), 30.

Treherne, J. (2002) *The Galápagos Affair*. London: Pimlico.

Tudge, C. (1992) *Last Animals at the Zoo: How Mass Extinction can be Stopped*. Oxford: Oxford University Press.

Vagvolgyi, J. (1974) Pinta tortoise: rediscovered. *Pacific Discovery*, **27**(2), 21–3.

Valle, C. A. (1994) Is conservation just a matter for the elite? A Galapagueno's viewpoint. *Noticias de Galápagos*, **53**, 2.

Van Denburgh, J. (1914) The gigantic land tortoises of the Galápagos archipelago. Expedition of the California Academy of Sciences to the Galápagos Islands, 1905–1906. *Proceedings of the California Academy of Sciences Series 4*, **2**(1), 203–374.

van Wieren, S. J. (2005) Populations: re-introductions. In van Andel, J. and Aronson, J. (eds) *Restoration Ecology*. Oxford: Blackwell.

Vargas, E. (1997) Sea cucumber fishing boat captured. *Noticias de Galápagos*, **58**, 2.

Velíz, E. (1995) Letter to President Ballen, 4 September.

Vences, M. et al. (2003) Multiple overseas dispersal in amphibians. *Proceedings of the Royal Society – Series B*, **270**, 2435–42.

Vogel, G. (2001) Cloned gaur a short-lived success. *Science*, **291**(5503), 409.

Walpole, M. J. et al. (2001) Pricing policy for tourism in protected areas: lessons from Komodo National Park, Indonesia. *Conservation Biology*, **15**, 218–27.

Walpole, M. J. and Leader-Williams, N. (2002) Ecotourism and flagship species in conservation. *Biodiversity and Conservation*, **11**, 543–7.

Wasser, S. K. et al. (2004) Scat detection dogs in wildlife research and management: application to grizzly and black bears in the Yellowhead ecosystem, Alberta, Canada. *Canadian Journal of Zoology*, **82**, 475–92.

Wayne, R. K. (1996) Conservation genetics in the Canidae. In Avise, J. C. and Hamrick J. L. (eds) *Conservation Genetics: Case Histories from Nature* (pp. 75–118). New York: Chapman & Hall.

Wayne, R. K. et al. (1991) Conservation genetics of the endangered Isle Royale gray wolf. *Conservation Genetics*, **5**(1), 41–51.

Weber, D. (1971) Pinta, Galápagos: Une ile a sauver: Pinta, Galápagos: An island to save. *Biological Conservation*, **4**(1), 8–12.

Wedekind, C. et al. (1995) MHC-dependent mate preferences in humans. *Proceedings of the Royal Society – Series B*, **260**, 245–9.

White, D. (2005) The Galápagos: unique and at risk. *Planet Earth*, summer, 22–3.

White, K. L. et al. (1999) Establishment of pregnancy after the transfer of nuclear transfer embryos produced from the fusion of argali (*Ovis ammon*) nuclei into domestic sheep (*Ovis aries*) enucleated oocytes. *Cloning*, 1(1), 47–54.

Wikelski, M. (1995) Setting a world heritage ablaze – the 1994 fire in the Galápagos. *International Forest Fire News*, 13, 8–11.

Wildt, D. E. et al. (2003) Toward more effective reproductive science for conservation. In Holt, W. V. et al. (eds) *Reproductive Science and Integrated Conservation* (pp. 2–23). Cambridge: Cambridge University Press.

Wilmut, I. et al. (1997) Viable offspring derived from fetal and adult mammalian cells. *Nature*, 385, 811–13.

Wilmut, I. et al. (2002) Somatic cell nuclear transfer. *Nature*, 419, 583–6.

Wilson, D. S. et al. (2003) Estimating age of turtles from growth rings: a critical evaluation of the technique. *Herpetologica*, 59(2), 178–94.

Wilson, P. J. et al. (2000) DNA profiles of the eastern Canadian wolf and the red wolf provide evidence for a common evolutionary history independent of the gray wolf. *Canadian Journal of Zoology*, 78, 2156–66.

Woram, J. (2005) *Charles Darwin Slept Here: Tales of Human History at World's End*. New York: Rockville Press.

Wright, J. W. (1983) The evolution and biogeography and evolution of the lizards of the Galápagos archipelago: evolutionary genetics of *Phyllodactylus* and *Tropidurus* populations. In Bowman, R. I. et al. (eds) *Patterns of Evolution in Galápagos Organisms* (pp. 123–155). San Fransisco, California: American Association for the Advancement of Science.

Wright, J. W. (1984) The origin and evolution of lizards of the Galápagos Islands. *Terra*, 22(4), 21–7.

Wyles, J. S. and Sarich, V. M. (1984) Are the Galápagos iguanas older than the Galápagos? *Biological Journal of the Linnean Society*, 21, 177–86.

FIGURE ACKNOWLEDGEMENTS

The author and publisher are grateful to all those who gave permission to use copyright material. Here follows a list of sources for the figures and acknowledgement of copyright.

p. vi: © Macmillan Science; P.1: Darwin (1899?); P.2: Darwin (1899?); 1.1: Vagvolgyi (1974). © California Wild; 1.2: Gunther (1877). © Peter Pritchard/Chelonian Research Institute; 1.3: © Ole Hamann; 1.4: © CDRS Image Library; 2.1: Photo by Steve Maslowski. US Fish and Wildlife Service; 2.2: US Fish and Wildlife Service; 2.3: © Justin Gerlach/Nature Protection Trust of the Seychelles; 2.4: © Linda Cayot; 3.1: Beebe (1924); 3.2: Beebe (1924); 3.3: Datasets from Smith and Sandwell (1997) and Farr and Kobrick (2001), combined by Jasper Konter of the Scripps Institution of Oceanography; 3.4: Arrowsmith (1839). © John Woram; 3.5: Drawing by John Megahan from Bunce et al. (2004) ; 4.1: Drawing by Meredith Nugent. © Mary Evans Picture Library/Explorer Archives; 4.2: © Henry Nicholls; 4.3: Gunther (1877). © Peter Pritchard/Chelonian Research Institute; 4.4: Based on Figure 5 from Caccone et al. (2002). © Macmillan Science; 4.5: US National Archives and Records Administration; 4.6: US Naval Historical Center; 5.1: Grand Canyon National Park Museum Collection, Image 5137; 5.2: © Galapagos Conservation Trust; 5.3: Photo by Lee Karney. US Fish and Wildlife Service; 5.4: © Henry Nicholls; 6.1: © Manfred Altamirano; 6.2: © Claudio Sillero/EWCP (www.ethiopianwolf.org); 7.1: Harris (1744). © John Woram; 7.2: © Derek Green; 7.3: Photo by Chris Servheen. US Fish and Wildlife Service; 7.4: © Peter Pritchard/Chelonian Research Institute; 7.5: © Peter Pritchard/Chelonian Research Institute; 8.1: © Henry Nicholls; 8.2: Darwin (1890). © John Woram; 8.3: Plate 21 from Rothschild (1915). © Natural History Museum, London; 9.1: © Michel Milinkovitch; 9.2: Photo by Fred Sorenson. US Fish and Wildlife Service; 10.1: Photo by LuRay Parker. US Fish and Wildlife Service; 10.2: © Daniel Gist; 10.3: © Beatrix Schramm; 11.1: © Pasqualino Loi; 11.2: Figure 1 from Gurdon and Colman (1999). © Nature Publishing Group; 11.3: Perry (1988). © Margaret Perry/Roslin Institute, Edinburgh; 11.4: University of Idaho. Copyright © Phil Schofield; E.1: © Ole Hamann.

INDEX

Please help us protect the giant tortoises of Galapagos.

Conservation in Galapagos depends on the donations and support of people throughout the world. Without your help, projects, such as the tortoise breeding centre at the Charles Darwin Research Station, would not be able to continue protecting the Galapagos giant tortoises from extinction.

You can help by becoming a friend of Galapagos or by sending a donation to one of the independent charities set up to help protect and conserve the unique flora and fauna of the Galapagos.

If you would like to help or receive further information, please contact one of the friends of Galapagos organisations:

Galapagos Conservation Trust
5 Derby Street
London W1J 7AB
United Kingdom
Website: www.gct.org
Email: gct@gct.org
Tel: +44 (0) 207 629 5049

Galapagos Conservancy
407 North Washington Street
Suite 105, Falls Church
VA 22046 USA
Website: www.galapagos.org
Email: Darwin@galapagos.org
Tel: +1 703 538 6833

Friends of
Galapagos
Saving Darwin's treasured islands
GALAPAGOS CONSERVATION TRUST

For a list of other friends of Galapagos organisations throughout the world, please visit http://www.gct.org/fogos.html

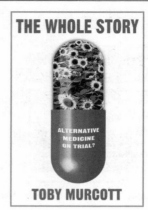
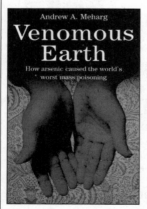

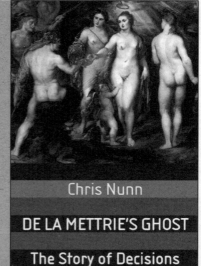